极客学院
jikexueyuan.com

互联网＋职业技能系列

职业入门 | **基础知识** | 系统进阶 | 专项提高

Android 移动应用开发基础教程

微课版

Android Development

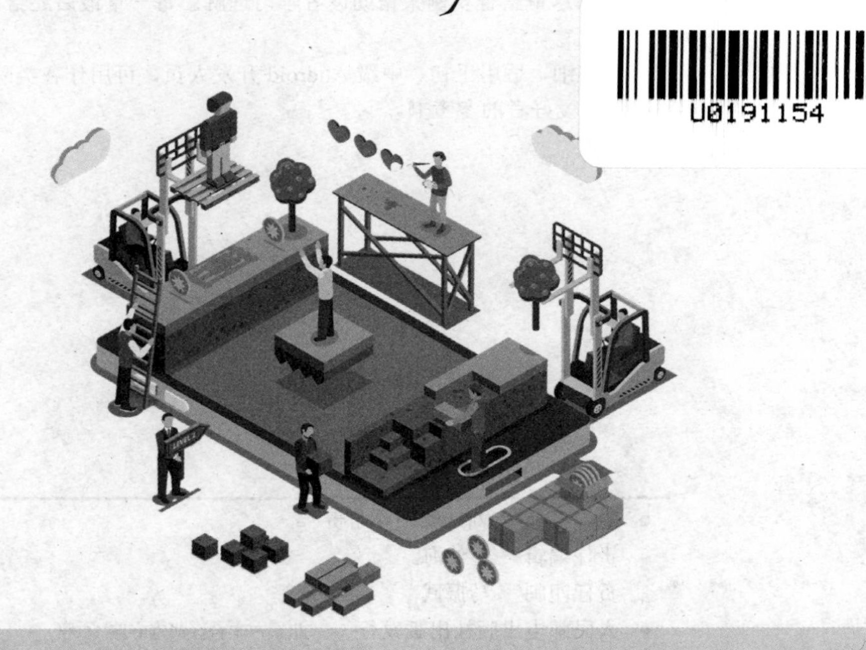

刘凡馨 夏帮贵 主编

人民邮电出版社
北京

图书在版编目（CIP）数据

Android移动应用开发基础教程 : 微课版 / 刘凡馨, 夏帮贵主编. -- 北京 : 人民邮电出版社, 2018.5（2021.6 重印）
（互联网+职业技能系列）
ISBN 978-7-115-47309-7

Ⅰ. ①A… Ⅱ. ①刘… ②夏… Ⅲ. ①移动终端－应用程序－程序设计－教材 Ⅳ. ①TN929.53

中国版本图书馆CIP数据核字(2017)第283672号

内 容 提 要

本书注重基础，循序渐进，系统地讲述了 Android 移动应用开发相关基础知识，涵盖了开发环境搭建、活动、UI 设计、广播机制、数据存储、多媒体、网络、数据解析、线程和服务等主要内容。对于每一个知识点，本书都尽量结合实例来帮助读者学习理解。每一章最后还给出一个综合实例来说明本章知识的使用。

本书内容丰富，讲解详细，适用于初、中级 Android 开发人员，可用作各类院校相关专业教材，同时也可作为 Android 开发爱好者的参考书。

◆ 主　　编　刘凡馨　夏帮贵
　责任编辑　左仲海
　责任印制　马振武

◆ 人民邮电出版社出版发行　　北京市丰台区成寿寺路 11 号
　邮编　100164　　电子邮件　315@ptpress.com.cn
　网址　http://www.ptpress.com.cn
　山东华立印务有限公司印刷

◆ 开本：787×1092　1/16
　印张：16.25　　　　　　2018 年 5 月第 1 版
　字数：479 千字　　　　　2021 年 6 月山东第 5 次印刷

定价：49.80 元

读者服务热线：(010) 81055256　印装质量热线：(010) 81055316
反盗版热线：(010) 81055315
广告经营许可证：京东市监广登字 20170147 号

前言
Foreword

Android系统的出现，使移动智能时代的发展进入了一个快速期，也使得移动终端厂商、移动系统企业的竞争加剧。Android手机、平板电脑、穿戴设备、车载设备越来越受用户欢迎，Android移动应用开发技术人员的需求也日益增大，Android程序设计已成为院校普遍开设的程序设计基础课程。

本书在内容编排和章节组织上特别针对院校教育特点，可让读者在短时间内掌握 Android 移动应用开发的基本方法。本书以“基础为主、实用为先、专业结合”为基本原则，在讲解 Android 移动应用开发基础知识的同时，力求结合项目实际，使读者能够理论联系实际，轻松掌握 Android 移动应用开发。

本书具有如下特点。

1. 入门条件低

读者无须太多技术基础，跟随教程即可轻松掌握Android移动应用开发的基本方法。

2. 学习成本低

本书在构建开发环境时，选择读者使用最为广泛的Windows操作系统和免费的Android Studio开发环境。

3. 内容编排精心设计

本书内容编排并不求全、求深，而是考虑读者的接受能力，选择必备、实用的 Android 相关知识进行讲解，各种知识和配套实例循序渐进、环环相扣，逐步涉及实际案例的各个方面。

4. 强调理论与实践结合

书中每个知识点都尽量安排一个短小、完整的实例，方便教师教学，也方便学生学习。

5. 丰富实用的课后习题

每章均准备一定数量的习题，方便教师安排作业，也方便学生通过练习巩固所学知识。

6. 精选极客学院在线课程

本书视频和相关实例均来自极客学院，同步讲解课程中的重点、难点。

7. 完整收集学习必备资源

读者可登录人民邮电出版社教育社区（http://www.ryjiaoyu.com/）下载书中所有实例代码、资源文件及习题参考答案。本书源代码可在学习过程中直接使用，参考相关章节进行配置即可。

本书作为教材使用时，课堂教学建议安排 40 学时，实验教学建议 24 学时。主要内容和学时安排如下表所示，教师可根据实际情况进行调整。

章	主要内容	课堂学时	实验学时
第 1 章	Android 开发起步	4	2
第 2 章	Android 核心组件——活动	6	4
第 3 章	UI 设计	8	4
第 4 章	广播机制	4	2
第 5 章	数据存储	6	4
第 6 章	多媒体	4	4
第 7 章	网络和数据解析	4	2
第 8 章	线程和服务	4	2

本书由西华大学刘凡馨、夏帮贵担任主编。刘凡馨编写第 1~4 章，夏帮贵编写第 5~8 章。刘凡馨负责全书统稿。

由于编者水平有限，书中难免存在疏漏之处，敬请广大读者批评指正。

编者

2017 年 7 月

目录
Contents

PART01

第1章

Android开发起步

重点知识：

- Android简介
- 搭建Android开发环境
- 创建第一个Android项目
- 了解Android项目组成
- 生成APK

■ Android 的横空出世，将智能设备的发展推向了一个新的快速发展时期。智能设备的普及发展，也使移动开发越来越受到开发者的青睐。移动操作系统目前分为三大领域，为 iOS、Android 和 Windows Phone，其中 Android 的发展最为迅猛，也最受人瞩目。本章通过介绍 Android、搭建开发环境、创建和认识 Android 项目，使读者对 Android 开发有一个初步了解。

1.1 Android 简介

Android 本义为“机器人”，现在人们看到它首先想到的是 Google 的移动操作系统。简单地说，Android 是基于 Linux 内核，应用 Java 开发的轻量级的移动操作系统。Google 为 Android 内置了诸多常用应用，如电话、短信、个人管理、多媒体播放、网页浏览等。

2003 年 10 月，Andy Rubin 等人创建了 Android 公司，组建了 Android 开发团队。2005 年 8 月，Google 收购了 Android 公司及其开发团队，并由 Andy Rubin 继续负责 Android 项目。2007 年 11 月，Google 正式发布 Android 平台，Android 平台也不再局限于手机，还逐渐扩展到平板电脑及其他智能设备。2011 年，Android 平台一举超过称霸移动领域多年的诺基亚 Symbian 系统，成为全球市场份额占有率第一的智能设备平台。

1.1.1 Android 平台特点

Android 平台具有下列主要特点。

1. 开放性

Android 平台源代码开放，开发人员可任意访问其核心代码，设计出更加丰富多彩的应用。Android 的开放性也使更多智能设备厂商加入到了 Android 联盟中来。

2. 不再受运营商限制

早期的手机上的应用、网络接入方式等，全部由运营商说了算，Android 打破了这种束缚，用户可以根据自己的喜好来定制手机应用。

3. 丰富的硬件选择

借助 Android 的开放性，硬件生产商可以设计出功能各异的多种产品，例如 Android 手机、平板电脑、眼镜、电视、车载设备以及穿戴设备等，为用户提供更多的选择。

4. 开发不受限制

Android 平台为开发人员提供了更加宽泛、自由的开发环境，使得各种优秀的应用不断出现。同时，这也导致了一些不健康、恶意应用的出现，如何遏制不良应用也成为 Android 面临的一个难题。

5. 与 Google 应用无缝结合

Android 平台可与 Google 地图、邮件、搜索等优秀服务无缝结合，在手机、平板电脑以及其他智能设备上可以轻松使用这些服务。

1.1.2 Android 平台体系架构

Android 平台体系架构如图 1-1 所示，从下到上依次分为 Linux 内核层（Linux Kernel）、硬件抽象层（Hardware Abstraction Layer（HAL））、系统运行库层（Native C/C++ Libraries 和 Android Runtime）、Java API 框架层（Java API Framework）和系统应用层（System Apps）。

1. Linux 内核层

Android 系统运行于 Linux 内核之上，Linux 内核层主要包括电源管理和各种启动模块，如显示驱动、键盘驱动、摄像头驱动、Wi-Fi 驱动及 USB 驱动等。

2. 硬件抽象层

硬件抽象层包含多个库模块，为 Java API 提供标准的设备硬件功能支持。开发人员通过框架 API 访问设备硬件时，Android 系统为硬件加载相应的库模块。

3. 系统运行库层

系统运行库层包含了一系列原生 C/C++库，它们通过 Android 应用框架 API 为开发者提供各种服务。例如，Webkit 库提供浏览器支持，OpenGL ES 库提供 2D/3D 绘画支持，等等。

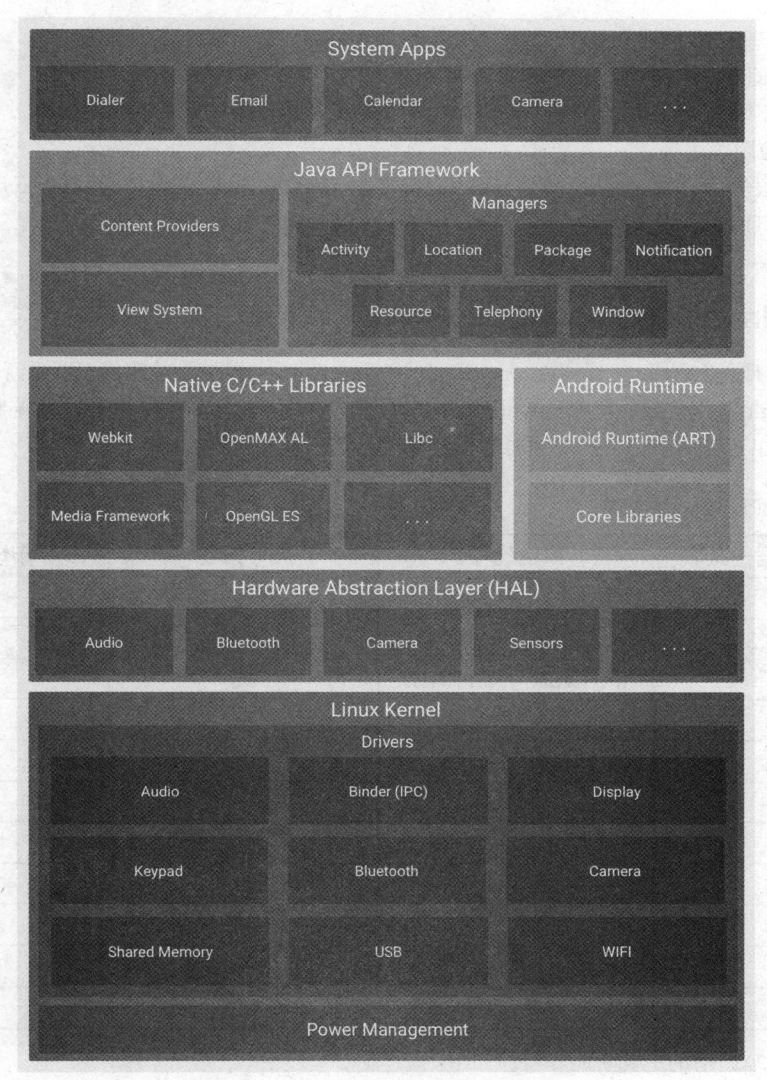

图 1-1　Android 平台体系架构

该层还包含了两个 Android 运行时库：核心库（Core Libraries）和 Android 运行时（Android Runtime，ART）。核心库允许开发人员使用 Java 开发 Android 应用。运行 Android 5.0（API 级别 21）及更高版本的设备，每个应用运行在自己的进程中，拥有自己的 ART 实例。ART 的主要功能包括预先和即时编译、优化的垃圾回收以及更好的调试支持等。

在 Android 5.0 之前，Android 运行时为 Dalvik 虚拟机。Dalvik 虚拟机是为移动设备定制的，针对手机内存、CPU 性能有限等特点做了专门的优化。不管是 ART 还是 Dalvik 虚拟机，Android 应用都被编译为 DEX 文件来执行。DEX 文件是专门为 Android 设计的字节码文件，占用内存更少，运行更快。

4. Java API 框架层

Java API 框架层通过 API 提供 Android 系统的全部功能，主要功能如下。

- 内容提供程序：为应用提供数据。
- 视图系统：提供应用 UI 设计支持，包括文本框、按钮、列表、可嵌入的网络浏览器等。

- 资源管理器：提供非代码资源管理功能，如字符串、图片和布局文件等。
- 通知管理器：为应用提供自定义状态栏通知功能支持。
- Activity 管理器：用于管理应用中活动的生命周期，提供导航返回栈。

5. 系统应用层

系统应用层包含了 Android 系统自带的一套核心应用，包括电子邮件、短信、日历、联系人等。系统自带应用可以为开发者提供功能支持，例如，调用系统自带的短信应用，可在第三方应用中实现短信发送功能。

1.1.3 Android 版本

2008 年 9 月，Google 发布了 Android 1.0。随后，Google 以惊人的速度不断更新 Android 系统。2016 年，Google I/O 大会上发布了 Android 7.0 系统，这是目前最新的 Android 系统版本。

表 1-1 列出了目前市场上一些主要的 Android 系统版本。

表 1-1 Android 系统主要版本

版本号	系统代号	API 级别	市场占有率
2.2	Froyo	8	0.1%
2.3.3 -2.3.7	Gingerbread	10	1.7%
4.0.3 -4.0.4	Ice Cream Sandwich	15	1.6%
4.1.x	Jelly Bean	16	6.0%
4.2.x	Jelly Bean	17	8.3%
4.3	Jelly Bean	18	2.4%
4.4	KitKat	19	29.2%
5.0	Lollipop	21	14.1%
5.1	Lollipop	22	21.4%
6.0	Marshmallow	23	15.2%
7.0	Nougat	24	<0.1%

Android 7.0 新增的主要功能和特性如下。

1. 多窗口支持

多窗口支持功能使用户可在运行 Android 7.0 系统的设备（手机、平板电脑或 TV）上一次打开两个应用。在 Android 7.0 手机和平板电脑中，用户可以并排运行两个应用，或者基于分屏模式应用两个应用。用户可拖动两个应用之间的分隔线来调整应用。在 Android 7.0 TV 中，同时运行的两个应用实现了画中画模式，在看电视的同时允许用户浏览或使用其他应用。

多窗口支持功能也允许在两个应用之间执行拖放操作，进一步增强了用户体验。

2. 通知功能增强

Android 7.0 重新设计了通知功能，使其速度更快，也更易于使用，主要改进如下所述。

- 模板进行了更新：模板更新后，开发人员只需修改少量代码即可实现通知。
- 允许更多自定义消息传递样式：使用 MessagingStyle 类的通知时，可自定义更多与通知有关的用户界面标签，可配置消息、会话标题和内容视图等内容。
- 捆绑通知：系统可将消息成组显示，用户可适当地进行拒绝或归档操作。
- 直接回复：在实时通信应用中支持内联回复，可以方便用户在通知界面中快速回复短信。
- 自定义视图：新的 API 允许在通知中使用自定义视图时充分利用系统装饰元素。

3. 及时编译(JIT)和预编译(AOT)

Android 7.0 添加了 JIT 编译器，可对 ART 进行代码分析，提升了应用性能。JIT 编译器对 AOT(Ahead of Time)编译器进行了补充，有助于提高运行性能，节约存储空间，加快应用和系统的更新速度。

配置文件指导的编译可让Android运行组件根据应用运行的实际情况管理每个应用的AOT/JIT编译。配置文件还可进一步指导编译减少内存占用，这对低内存设备尤其重要。通过配置文件指导，还可在设备处于空闲或充电状态时进行编译，从而节约时间并省电。

4. 随时随地的低耗电模式

Android 6.0 推出了低耗电模式，当设备未连接电源、处于静止状态且屏幕关闭时，设备自动进入低耗电模式，系统通过推迟应用的 CPU 和网络活动来达到省电的目的。

Android 7.0 进一步完善了低耗电模式。只要屏幕关闭且未连接电源，但不一定处于静止状态(例如用户将手机放于口袋中)，低耗电模式就会启动，限制 CPU 和网络活动。

5. 流量节省程序

Android 7.0 推出了流量节省模式，允许用户启用流量节省程序，当设备使用按流量计费的网络时，系统可屏蔽后台流量，同时指示前台应用尽可能少用流量。例如限制流媒体服务的比特率，降低图像质量，延迟最佳的预缓冲等。用户还可将应用加入白名单，从而允许其在后台访问网络。

6. 号码屏蔽

Android 7.0 增加了号码屏蔽功能，允许默认短信应用、默认手机应用和运营商应用通过框架 API 访问屏蔽的号码列表，而其他应用无法访问此列表。利用平台标准的号码屏蔽功能，可以屏蔽已屏蔽号码发出的短信，可通过备份/还原重置或跨设备保留屏蔽的号码，可在多个应用中使用相同的屏蔽号码列表。Android 设备的运营商可通过读取用户设备中的屏蔽号码列表，执行服务器端的屏蔽，阻止已屏蔽号码的来电和短信到达用户设备。

1.2 搭建 Android 开发环境

“工欲善其事，必先利其器”，选择一款好的开发工具，有助于提高开发效率。本节将介绍如何搭建 Android 开发环境。

1.2.1 需要哪些工具

Android 开发需要的工具如下所述。

- JDK：Android 程序都使用 Java 语言进行编写，JDK 是 Java 语言开发工具包，它包含了 Java 运行环境、工具、基础类库等。目前，Android 支持 Java 7 的全部功能和 Java 8 的部分功能。
- Android SDK：这是 Google 提供的 Android 开发工具包，开发 Android 应用时，需要在集成开发环境(IDE)中引入该包。
- Android Studio：这是 Google 推出的 Android 开发 IDE。早期的 Android 开发大多使用 Eclipse，在其中安装了 Google 提供的 Android 开发插件 ADT。随着 Android Studio 的不断完善和功能的不断增强，Android Studio 逐渐成为 Android 开发的理想选择，所以 Google 也不再维护和更新 ADT 插件。

对于上述工具，应根据所使用的操作系统类型和版本安装对应的版本，本书将在 32 位的 Windows 10 操作系统中完成所有开发。

1.2.2 JDK 下载安装

Windows 下 JDK 的下载与安装

本书将使用 JDK 8 进行 Android 开发。读者要使用本书提供的源代码，则需安装 JDK 8 或更高版本。

JDK 8 的下载安装步骤如下所述。

（1）查看操作系统信息。在 Windows 任务栏中右击 Windows 图标，然后在弹出的菜单中选择“系统”命令，打开 Windows 系统信息窗口，如图 1-2 所示。

图 1-2　Windows 操作系统信息窗口

（2）记住 Windows 系统信息窗口中显示的操作系统类型、版本，在后续的操作中要选择对应的 JDK。

（3）打开浏览器，在地址栏中输入 JDK 下载网站地址：http://www.oracle.com/technetwork/java/javase/downloads/jdk8-downloads-2133151.html，打开下载页面，如图 1-3 所示。

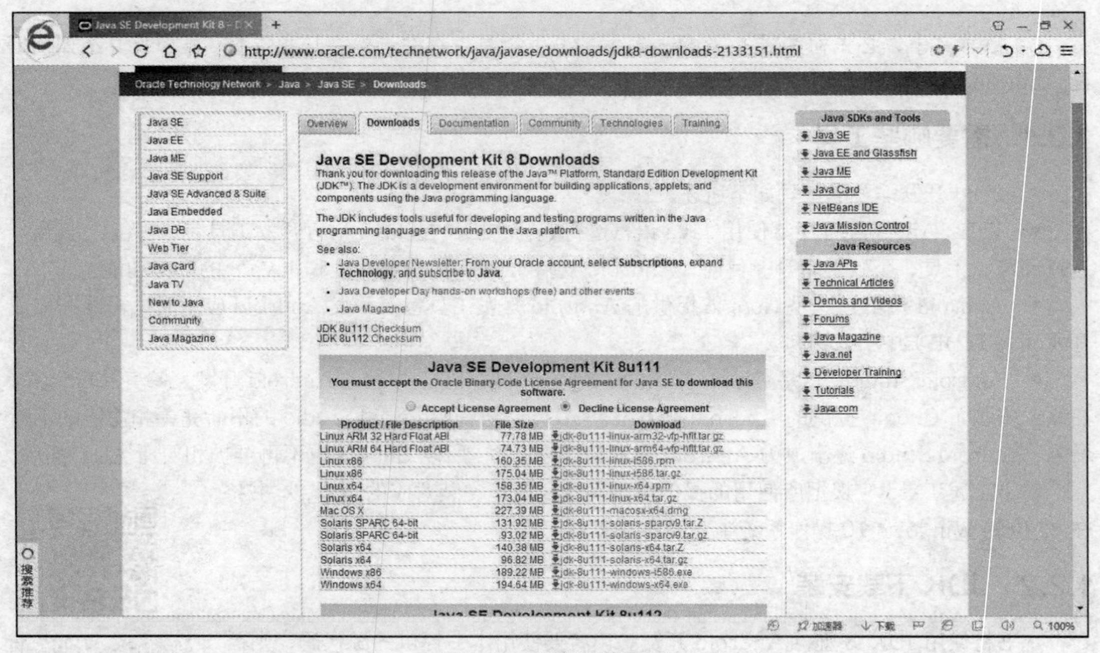

Product / File Description	File Size	Download
Linux ARM 32 Hard Float ABI	77.78 MB	jdk-8u111-linux-arm32-vfp-hflt.tar.gz
Linux ARM 64 Hard Float ABI	74.73 MB	jdk-8u111-linux-arm64-vfp-hflt.tar.gz
Linux x86	160.35 MB	jdk-8u111-linux-i586.rpm
Linux x86	175.04 MB	jdk-8u111-linux-i586.tar.gz
Linux x64	158.35 MB	jdk-8u111-linux-x64.rpm
Linux x64	173.04 MB	jdk-8u111-linux-x64.tar.gz
Mac OS X	227.39 MB	jdk-8u111-macosx-x64.dmg
Solaris SPARC 64-bit	131.92 MB	jdk-8u111-solaris-sparcv9.tar.Z
Solaris SPARC 64-bit	93.02 MB	jdk-8u111-solaris-sparcv9.tar.gz
Solaris x64	140.38 MB	jdk-8u111-solaris-x64.tar.Z
Solaris x64	96.82 MB	jdk-8u111-solaris-x64.tar.gz
Windows x86	189.22 MB	jdk-8u111-windows-i586.exe
Windows x64	194.64 MB	jdk-8u111-windows-x64.exe

图 1-3　JDK 下载页面

（4）选中 Accept License Agreement 选项，然后根据使用的操作系统选择下载安装文件。本书使用 32 位的 Windows 10 操作系统，所以选择并下载 Windows x86 安装文件，文件名为 jdk-8u111-windows-i586.exe。

（5）下载完成后，运行 jdk-8u111-windows-i586.exe，启动 JDK 安装程序，如图 1-4 所示。

（6）单击 下一步(N) > 按钮，打开定制安装界面，如图 1-5 所示。

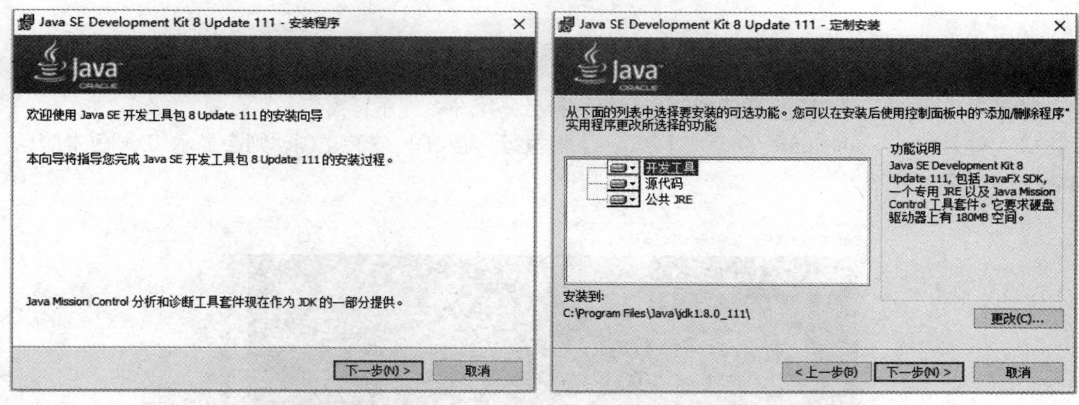

图 1-4　JDK 安装程序　　图 1-5　定制安装

（7）在定制安装界面中选择安装的组件和安装位置。默认情况下会安装全部组件，包括开发工具、源代码和公共 JRE，其中只有“开发工具”是必需的。若要取消某一项组件，在列表中单击组件前面的图标，在弹出的菜单中选择“此功能不可用”命令即可。本书中，JDK 默认安装到 C:\Program Files\Java\jdk1.8.0_111\，单击 更改(C)... 按钮可改变安装路径。最后，单击 下一步(N) > 按钮，安装程序根据设置执行安装过程。

（8）若选择了安装公共 JRE，安装程序会打开图 1-6 所示的界面，提示选择公共 JRE 的安装位置。单击 更改(C)... 按钮可更改安装位置，最后单击 下一步(N) > 按钮执行公共 JRE 安装即可。

（9）安装完成后，显示如图 1-7 所示的结束界面，单击 关闭(C) 按钮结束安装。

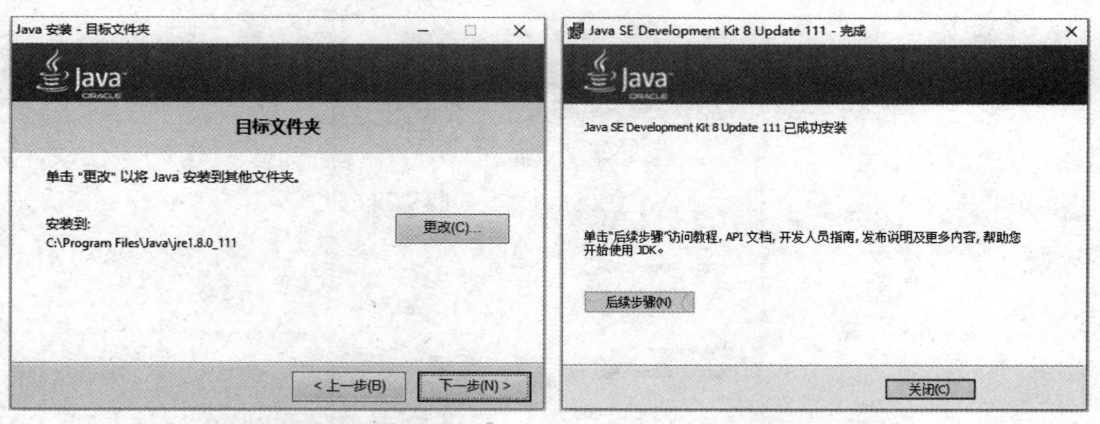

图 1-6　指定公共 JRE 安装位置　　图 1-7　安装完成

JDK 安装完成后，可按照下面的步骤测试 JDK 是否安装正确。

（1）在 Windows 任务栏中右击 Windows 图标，然后在弹出的菜单中选择“运行”命令，打开“运行”对话框，如图 1-8 所示。

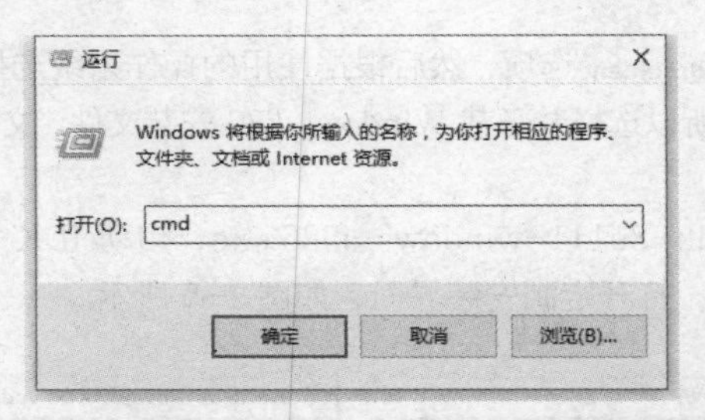

图 1-8 “运行”对话框

（2）输入 cmd，单击确定按钮执行命令，打开系统命令提示符窗口。

（3）输入 java -version 命令，按【Enter】键执行。若 JDK 安装正确，则会显示 JDK 版本信息，如图 1-9 所示。

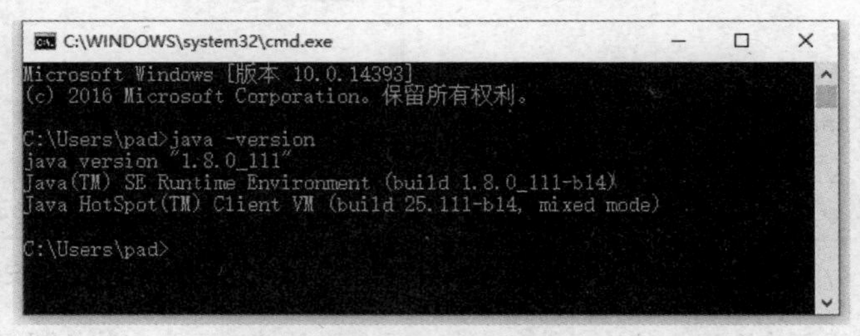

图 1-9 查看 JDK 版本信息

在完成 JDK 安装后，如果 Android Studio 找不到 JDK，则需要为 Windows 系统添加 JAVA_HOME 环境变量，指定 JDK 安装路径（注意这里指完整的 JDK 安装路径，不是 JRE 安装路径）。

配置 Windows 系统环境变量的操作步骤如下。

（1）在 Windows 任务栏中右击 Windows 图标，然后在弹出的菜单中选择“系统”命令，打开 Windows 系统信息窗口，如图 1-2 所示。

（2）单击 Windows 系统信息窗口左侧的“高级系统设置”选项，打开“系统属性”对话框，如图 1-10 所示。

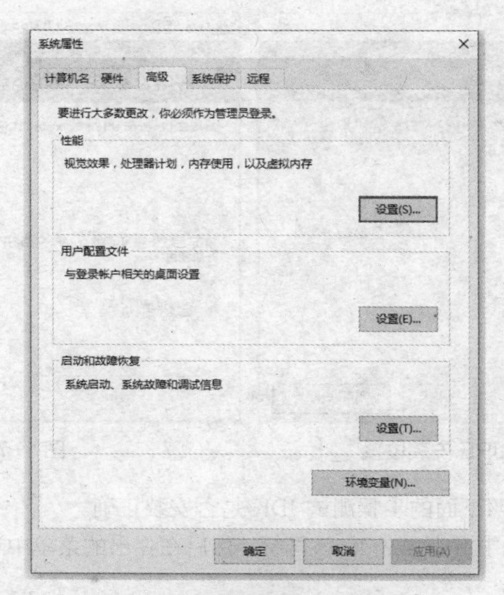

图 1-10 “系统属性”对话框

（3）单击“系统属性”对话框下方的 环境变量(N)... 按钮，打开“环境变量”对话框，如图 1-11 所示。“环境变量”对话框上面的列表框中显示了当前用户的环境变量，只对当前用户有效；对话框下面的列表框中显示了系统环境变量，对系统所有用户均有效。这里创建两种环境变量均可。

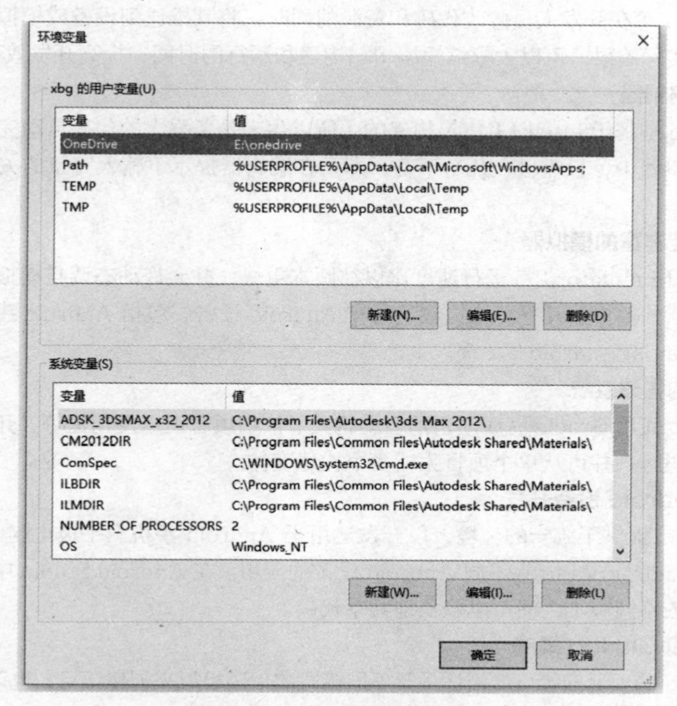

图 1-11 “环境变量”对话框

（4）单击“系统变量”列表栏下的 新建(N)... 按钮，打开“新建系统变量”对话框，如图 1-12 所示。

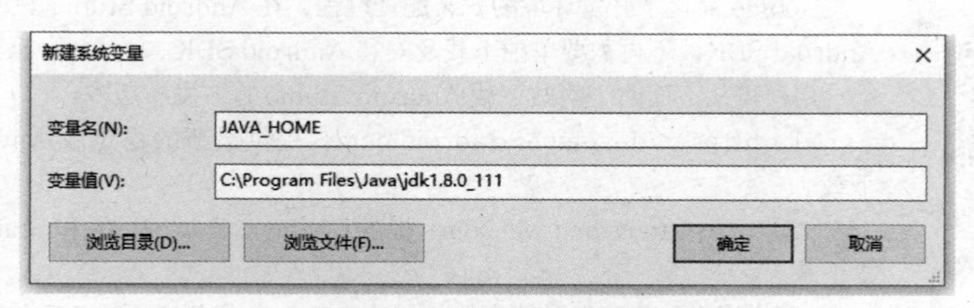

图 1-12 “新建系统变量”对话框

（5）在“新建系统变量”对话框的“变量名”文本框中输入 JAVA_HOME，在“变量值”文本框中输入 JDK 安装路径，如 C:\Program Files\Java\jdk1.8.0_111。最后单击 确定 按钮关闭对话框，完成系统变量创建操作。

1.2.3 Android Studio 简介

Android Studio 简介

Android Studio 是 Google 推出的 Android 应用集成开发环境。在 2013 年的 Google I/O 大会上，Google 发布了 Android Studio 的第一个预览版本。Google 停止了 Eclipse 插件 ADT 的更新，并不断完善和更新 Android Studio，

截至 2016 年 12 月，最新的稳定版本为 Android Studio v2.2.3，2017 年 1 月最新版本为 Android Studio 2.3 Beta 2。

Android Studio 的主要特点如下。

1. Instant Run

Android Studio 可在开发人员修改代码和资源的同时，将其推送到设备或模拟器中，让开发人员立刻看到更改后的实际效果，可以大幅缩短编辑、构建和运行的时间，提高开发效率。

2. 智能代码编辑器

Android Studio 是基于 IntelliJ IDEA 构建的，所以生来具备强大的代码自动完成、重构和代码分析功能。在代码编辑器中，只需输入几个字符，即可在编辑器提示下输入完整的关键词，快速完成代码编辑。

3. 快速、功能丰富的模拟器

最新的 Android Emulator 2.0 运行速度比以往版本更快，并允许动态调整模拟器的大小以及访问一组传感器控件。新的模拟器几乎可模拟测试所有 Android 设备，包括 Android 手机、Android 平板电脑、Android Wear 和 Android TV 等。

4. 强大灵活的构建系统

Android Studio 使用 Gradle 完成项目的自动构建、依赖项管理以及构建等，开发人员可轻松地将项目配置为包含代码库，并可从单个项目生成多个构建变体。

5. 专用于 Android 设备开发

Android Studio 提供了统一的环境，以开发适用于 Android 手机、平板电脑、Android Wear、Android TV 以及 Android Auto 等不同 Android 设备的应用。在 Android Studio 中，单个项目可应用于多种设备，应用的不同版本之间可以轻松地共享代码。

6. 代码模板和 GitHub 集成

Android Studio 的项目向导可使用适用于不同模式的代码模板来创建项目，也可以从 GitHub 导入 Google 代码示例，使开始一个项目变得十分简单。

1.2.4 Android Studio 的下载安装

Android Studio
下载安装

Google 简化了开发环境的安装配置过程，在 Android Studio 安装包中集成了 Android SDK，不再需要单独下载及安装 Android SDK。

用户可从下面两个网站下载 Android Studio 官方发布版本。

（1）https://developer.android.com：提供最新的稳定版 Android Studio 下载。

（2）http://tools.android.com：提供 Canary、Dev、Beta 和 Stable 等 4 种版本下载。

- Canary：Canary 通道提供处于开发过程中的 Android Studio 最新版本，几乎每周更新。
- Dev：经过一轮完整内部测试的 Canary 版本会进入 Dev 通道。
- Beta：稳定的 Canary 版本会进入 Beta 通道，作为正式发布的稳定版的候选版本。
- Stable：Stable 通道提供正式发布的稳定版本，最新的稳定版与 https://developer.android.com 中提供的保持一致。

Android Studio 下载安装步骤如下所述。

（1）在浏览器地址栏中输入 http://tools.android.com/download/studio，打开下载导航页面，如图 1-13 所示。

（2）单击页面中的 Stable 超链接，进入稳定版下载导航页面，如图 1-14 所示。

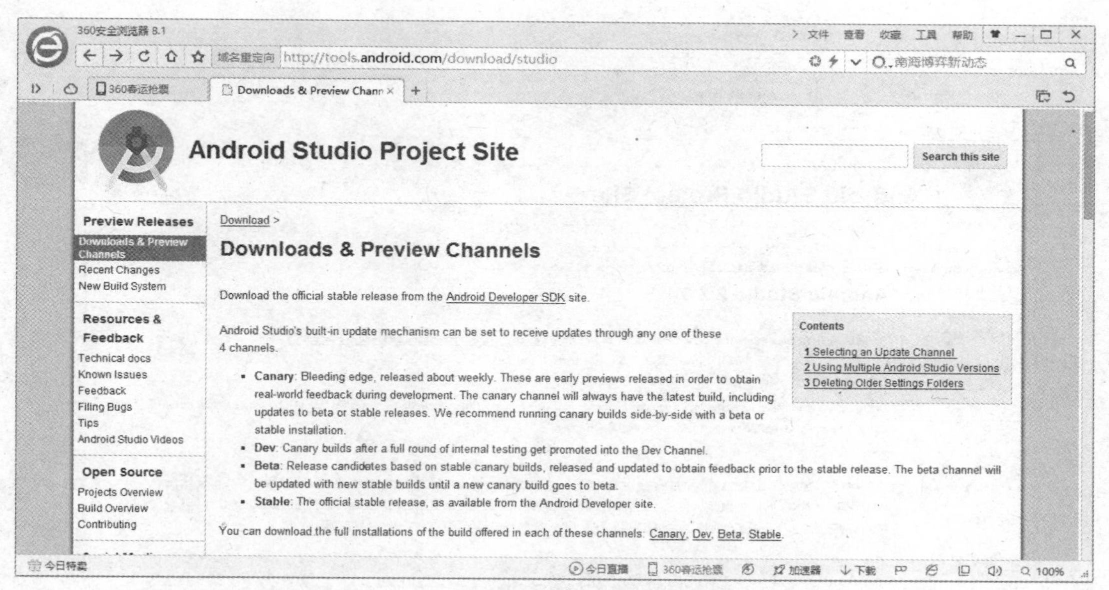

图 1-13　下载导航页面

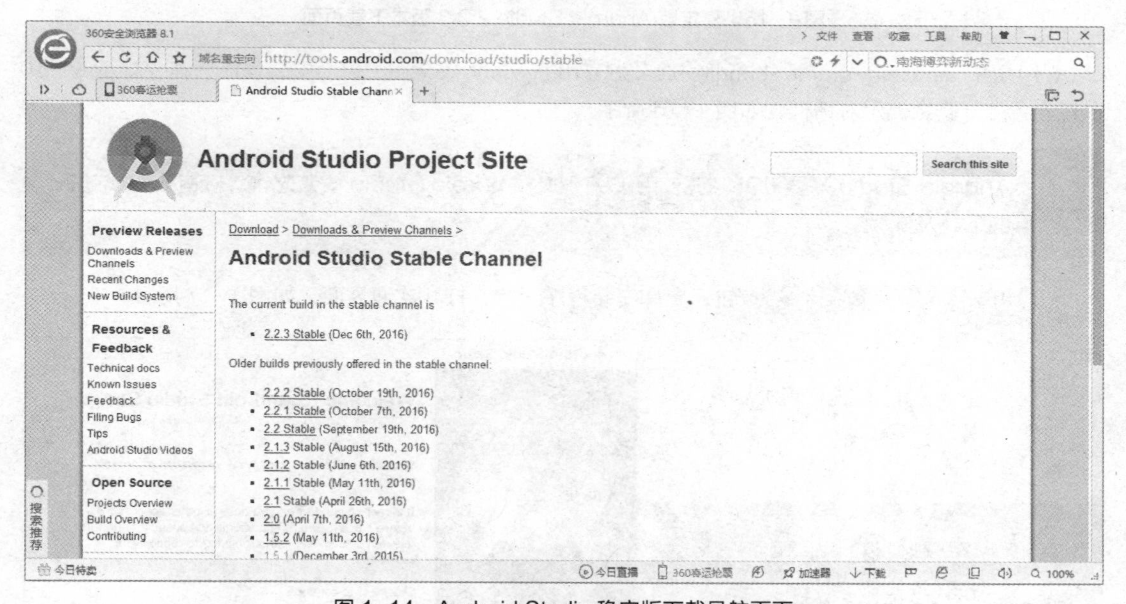

图 1-14　Android Studio 稳定版下载导航页面

（3）这里可看到最新的稳定版为 2016 年 12 月的 2.2.3 版本。单击“2.2.3 Stable”超链接，进入下载页面，如图 1-15 所示。页面中提供了 3 种安装程序（Installers）和 4 种压缩文件（Zip Files）供用户下载，建议 Windows 系统用户下载“Windows bundle with SDK”安装程序，其中包含了 Android Studio 安装程序和 Android SDK。单击超链接 https://dl.google.com/dl/android/studio/install/2.2.3.0/android-studio-bundle-145.3537739-windows.exe 即可开始下载。

在 https://developer.android.com/studio/中也可直接下载绑定了 Android SDK 的最新稳定版的 Android Studio 安装程序。

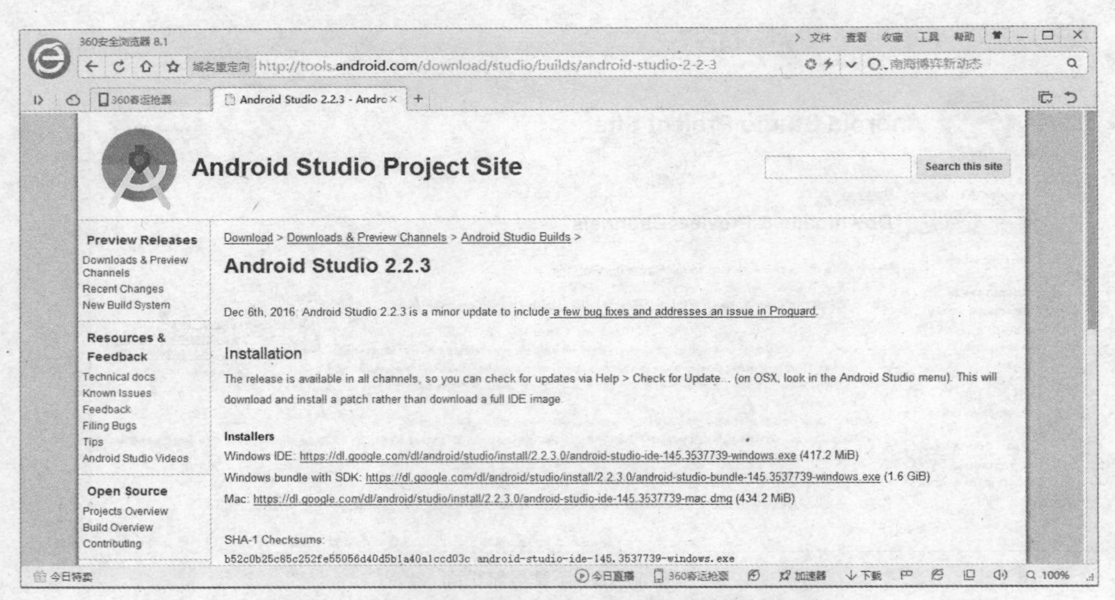

图 1-15　稳定版 Android Studio 2.2.3 版本下载页面

（4）运行 android-studio-bundle-145.3537739-windows.exe，启动安装程序，在 Windows 10 中首先会打开安全提示对话框，如图 1-16 所示。

Android Studio 需要 JDK 支持，所以在开始 Android Studio 安装之前，应首先完成 JDK 的下载及安装。

（5）单击 是 按钮，允许安装程序运行，打开欢迎界面，如图 1-17 所示。

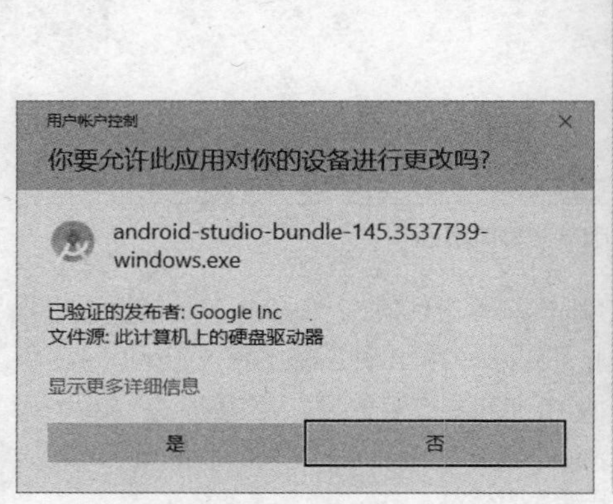

图 1-16　安全提示对话框

图 1-17　Android Studio 安装程序欢迎界面

（6）单击 Next > 按钮，打开选择组件界面，如图 1-18 所示。

（7）默认情况下安装 Android Studio、Android SDK 和 Android Virtual Device 全部组件。接受默认选择，单击 Next > 按钮，打开软件协议界面，如图 1-19 所示。

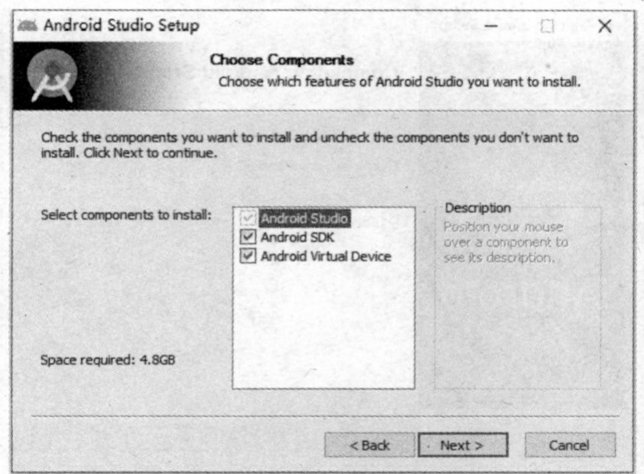

图 1-18　选择安装组件

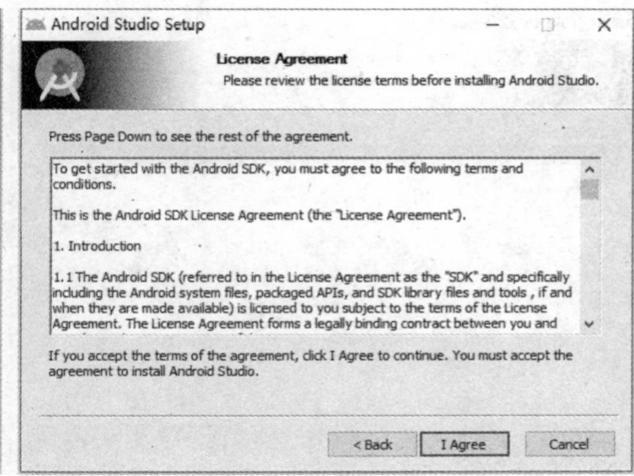

图 1-19　选择软件协议

（8）单击 I Agree 按钮，同意软件协议，打开配置选项界面，如图 1-20 所示。

（9）在配置选项界面中可设置 Android Studio 和 Android SDK 的安装路径。Android Studio 需要 500MB 左右的磁盘空间，默认安装到系统盘的\Program Files\Android\Android Studio 文件夹中。Android SDK 需要 3.2GB 左右的磁盘空间，默认安装到系统盘的\Users\xbg\AppData\Local\Android\sdk 文件夹。对于系统盘空间有限的用户，建议将 Android SDK 安装到非系统盘。最后，单击 Next > 按钮，打开选择开始菜单文件夹界面，如图 1-21 所示。

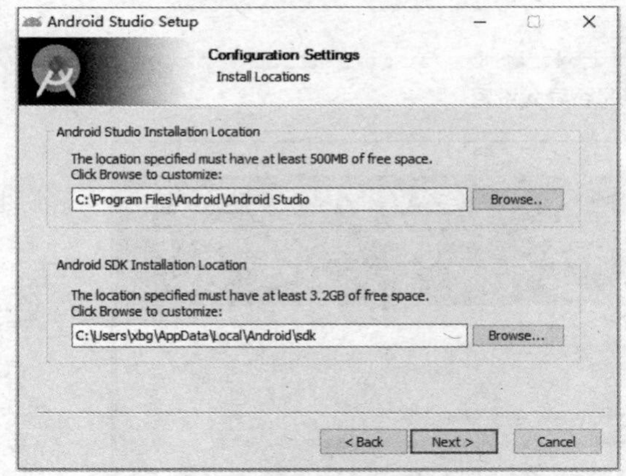

图 1-20　配置安装目录

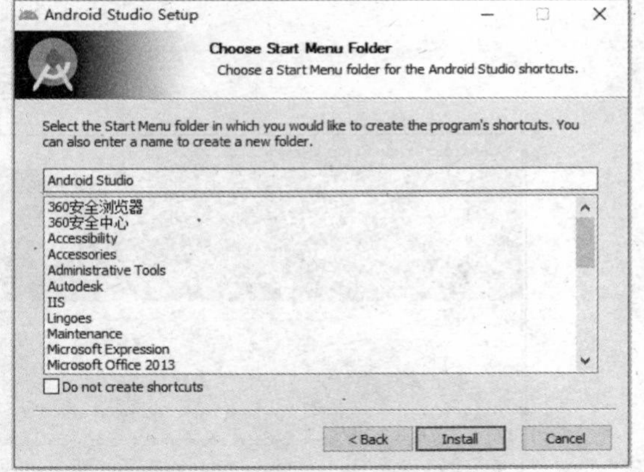

图 1-21　选择开始菜单文件夹

（10）在选择开始菜单文件夹界面中修改 Android Studio 在 Windows 开始菜单中的文件夹名称，默认为 Android Studio。单击 Install 按钮，开始执行安装操作。

（11）文件复制完成后会打开图 1-22 所示的安装完成界面，单击 Next > 按钮，打开完成 Android Studio 安装界面，如图 1-23 所示。

（12）在完成 Android Studio 安装界面中默认勾选了 Start Android Studio 复选框，表示用户可在完成安装后自动启动 Android Studio。最后，单击 Finish 按钮结束安装。

在安装完成后，如果选择了启动 Android Studio，则会打开如图 1-24 所示的对话框，选中 I want to import my settings from a custom location 单选按钮，表示可从指定位置导入早期版本的 Android Studio 设置；选中 I do not have a previous version of Studio or I do not want to import my settings 单选按钮，表示没有安装其他版本的 Android Studio 或不想导入以前的设置；单击 OK 按钮即可启动 Android Studio，打开设置向导的欢迎界面，如图 1-25 所示。第一次启动时才会打开设置向导，可验证当前安装的 Android SDK 和开发环境，并可从网络下载需要的相关组件。

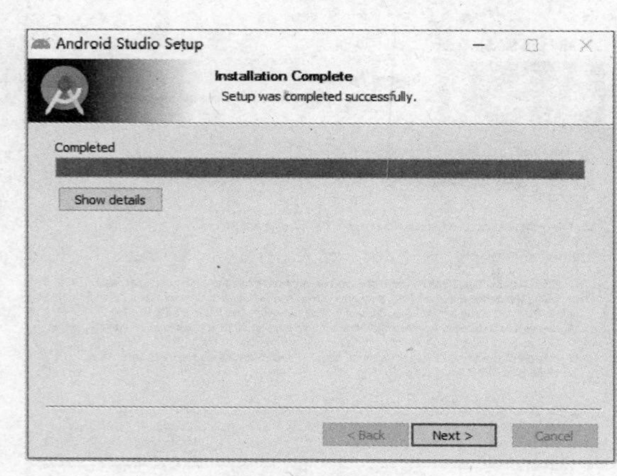

图 1-22　安装完成

图 1-23　完成 Android Studio 安装

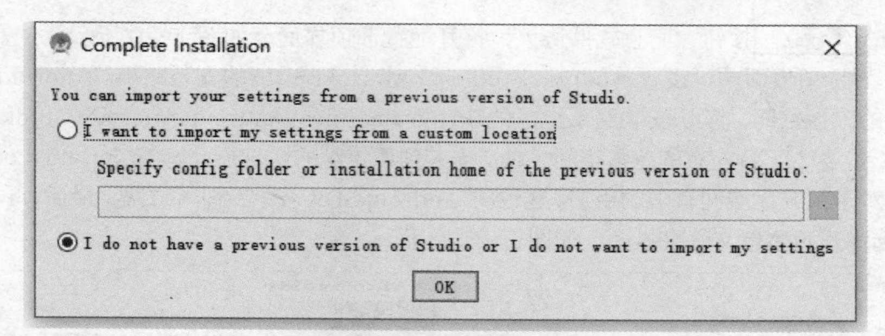

图 1-24　选择是否导入设置

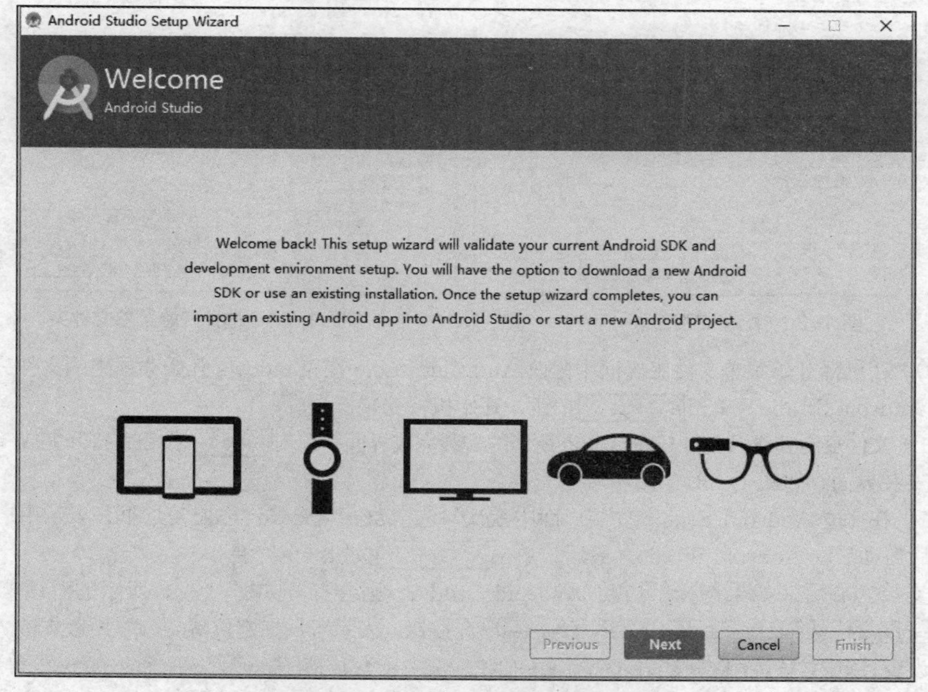

图 1-25　设置向导的欢迎界面

单击 Next 按钮，设置安装类型，如图 1-26 所示。

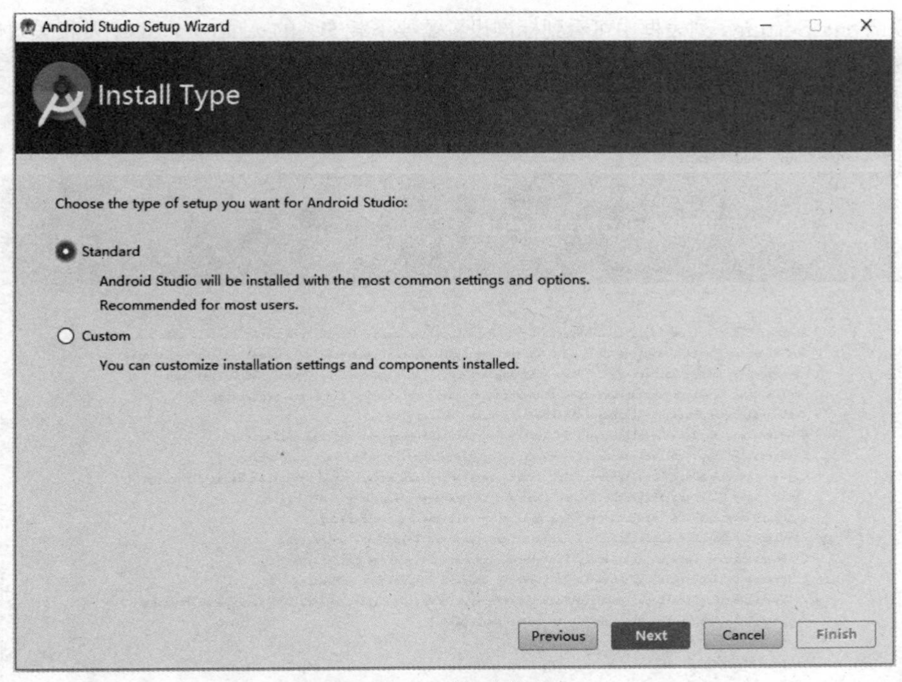

图 1-26　选择安装类型

在安装类型界面中选择如何设置 Android Studio，选中 Standard 单选按钮表示按照标准方式设置，选中 Custom 单选按钮表示自定义设置。通常选择标准设置，选择自定义则可详细了解设置向导的操作细节。

选中 Standard 单选按钮，单击 Next 按钮，打开确认设置对话框，如图 1-27 所示。

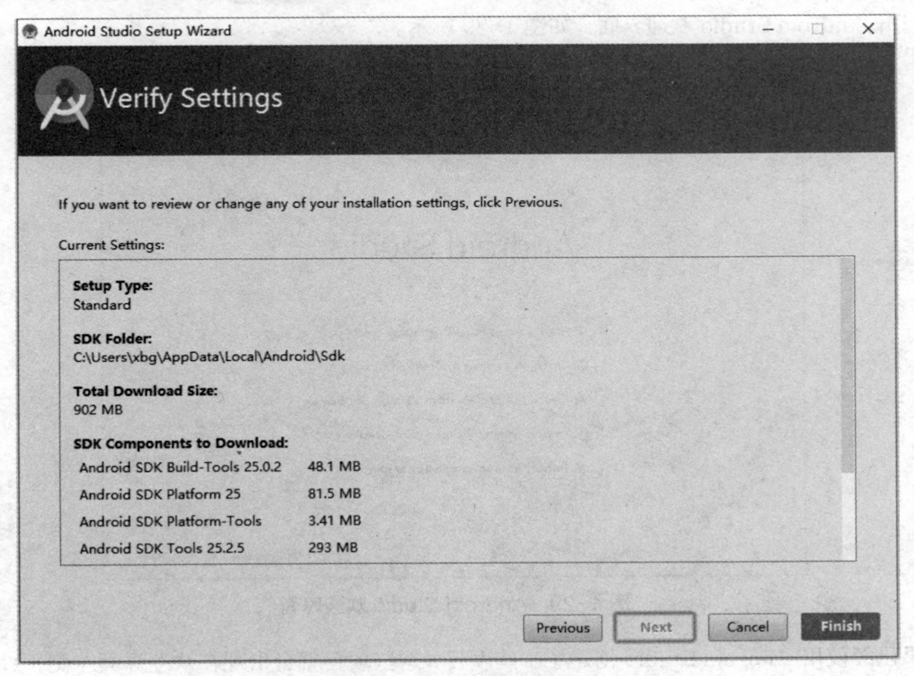

图 1-27　确认设置

在 Current Settings 列表框中可看到标准设置需要下载安装的 Android SDK 及相关组件，单击 Finish 按钮，Android Studio 将根据设置下载 Android SDK 相关组件。下载并安装完成后，会显示如图 1-28 所示的完成界面。

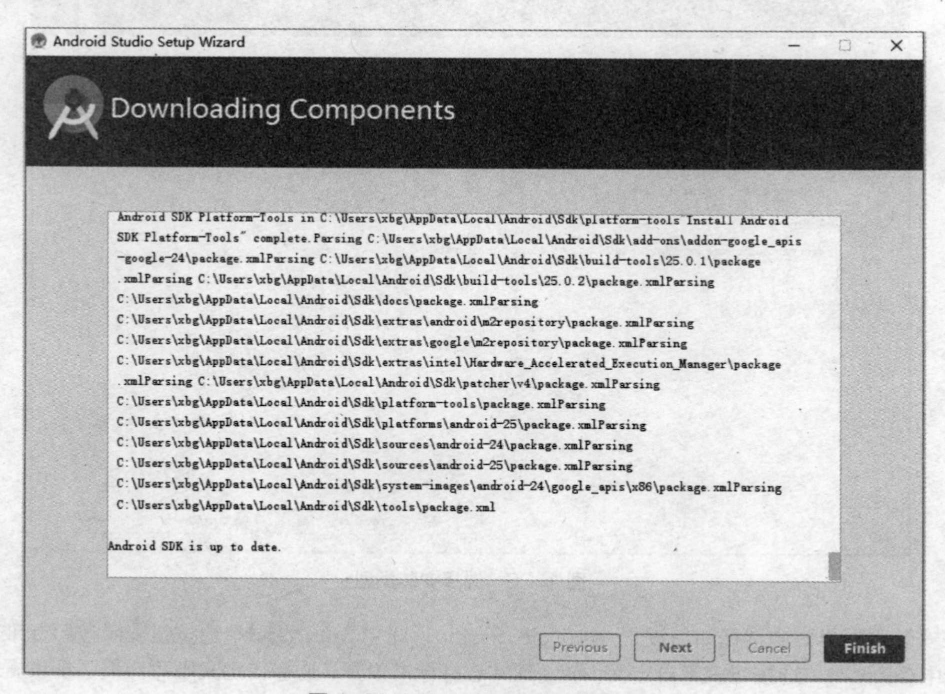

图 1-28 Android SDK 下载完成

在列表中可查看已经下载及安装的 Android SDK 组件。单击 Finish 按钮即可完成 Android Studio 设置，打开 Android Studio 欢迎界面，如图 1-29 所示。

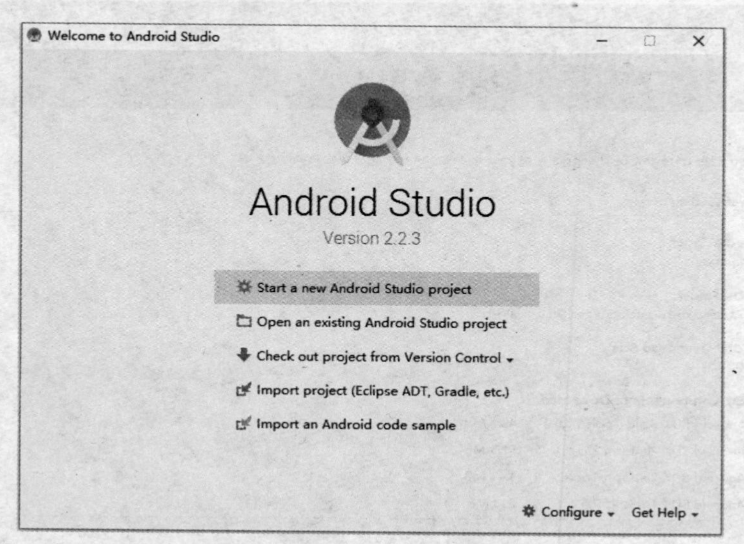

图 1-29 Android Studio 欢迎界面

如果已经使用 Android Studio 创建了一些项目，这些项目就会出现在欢迎界面左侧，如图 1-30 所示。

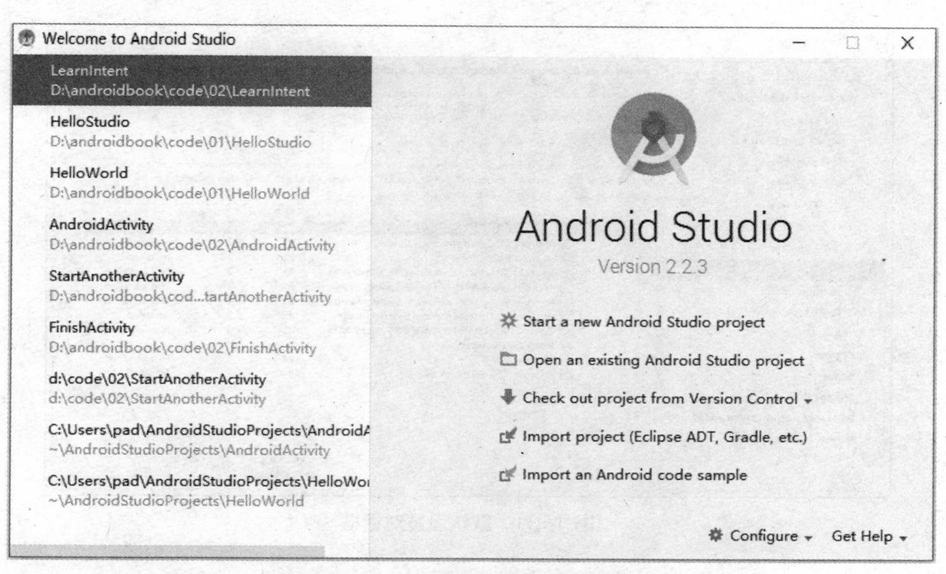

图 1-30　包含项目列表的欢迎界面

在欢迎界面中可完成下列操作。

- 创建新项目：单击“Start a new Android Studio project”选项，可打开新建项目向导，创建一个新的 Android Studio 项目。
- 打开现有项目：单击左侧的项目名称，可打开该项目。或者单击“Open an existing Android Studio project”选项，在打开的对话框中选择打开其他现有的项目。
- 通过版本控制系统检出项目：单击“Check out project from Version Control”选项，可从 GitHub、CVS、Google 云等多种版本控制系统中检出项目。
- 导入项目：单击“Import project (Eclipse ADT, Gradle, etc.)”选项，可将现有的 Eclipse ADT 项目、Gradle 项目等导入 Android Studio 中。
- 导入 Android 代码示例：单击“Import an Android code sample”选项，可导入 Android 示例代码来创建项目。
- 执行配置：从 Configure 菜单中可选择命令对 Android Studio 执行各种设置。例如打开 SDK 管理器，查看各个版本的 Android SDK 和相关组件是否安装，并安装没有的组件。
- 查看帮助：在 Get Help 菜单中，可选择命令查看 Android Studio 相关的帮助信息。

在欢迎界面中单击 Configure，打开选项菜单，选择“SDK Manager”命令，可打开默认设置对话框，并显示 Android SDK 平台清单，如图 1-31 所示。

在默认设置对话框中几乎可完成 Android Studio 相关的所有设置，在左侧列表中选中相关选项，即可在对话框右侧进行设置。这里只简单说明 Android SDK 平台清单的管理。勾选 Show Package Details 复选框，显示每个版本 Android SDK 的相关组件，勾选未安装组件名称前面的复选框，单击 Apply 按钮可下载并安装该组件；取消勾选安装组件名称前面的复选框，单击 Apply 按钮可删除已经安装的组件。

在 Android Studio 中创建 Android 应用时，可选择使用不同版本的 Android SDK。在运行应用时，如果使用的 SDK 还没有安装，Android Studio 会提示下载安装。使用 SDK 管理器，可提前将要使用的 SDK 下载并安装到本地计算机中，这样即使在没有网络连接时，Android SDK 也可使用。

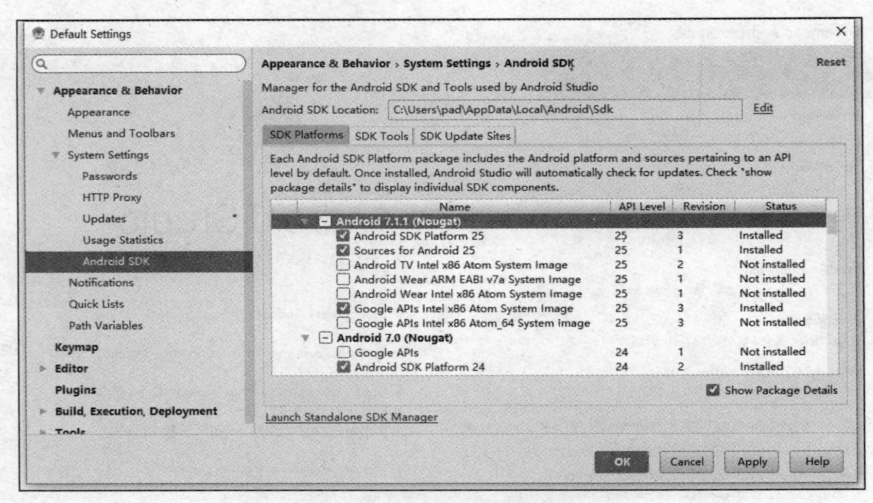

图 1-31　默认设置对话框

1.3　创建第一个 Android 项目

使用 Android Studio 可轻松完成项目的创建、界面设计、代码编写和测试运行等操作。

创建 HelloWorld 项目

1.3.1　创建 HelloWorld 项目

HelloWorld 项目运行时在屏幕上显示“Hello World”字符串，读者可以通过创建该项目了解如何使用 Android Studio 创建一个新的项目。

具体操作步骤如下。

（1）在 Android Studio 欢迎界面中单击“Start a new Android Studio project”选项，打开创建新项目对话框，如图 1-32 所示。

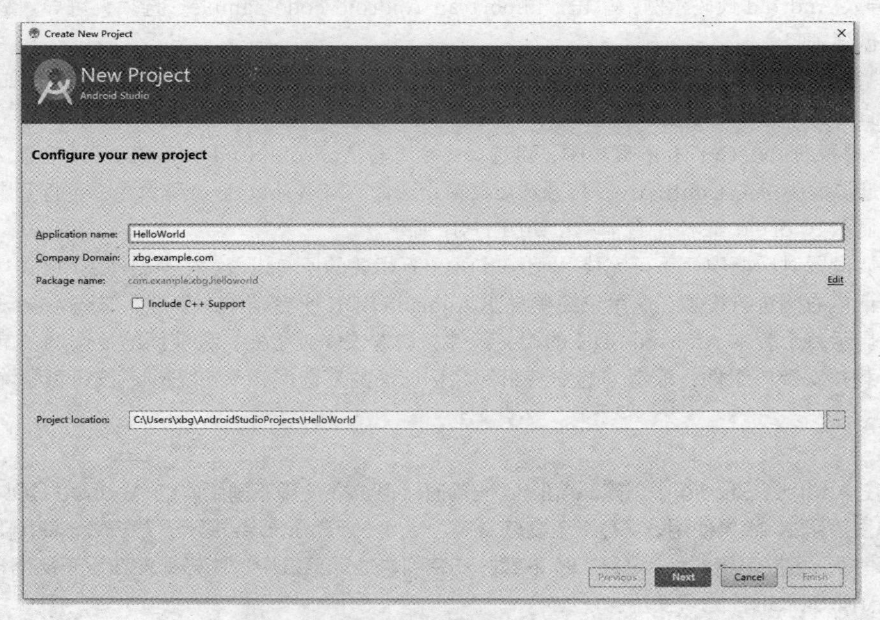

图 1-32　创建新项目

（2）在 Application name 文本框中输入 HelloWorld 作为应用名称，将应用安装到设备中时会显示该名称。在 Company Domain 文本框中输入公司域名，可任意输入一个域名或使用默认名称。Package name 为包的名称，Android 系统通过它来区分应用。Android Studio 会根据应用名称和公司域名自动生成一个包名。如果要修改包名，可单击右侧的 Edit 超链接，然后进行修改。若勾选 Include C++ Support 复选框，则可在项目中使用 C++。Project location 文本框显示了存放项目文件的默认文件夹，通过单击右侧的 ... 按钮可选择其他文件夹。最后，单击 Next 按钮，打开选择目标设备对话框，如图 1-33 所示。

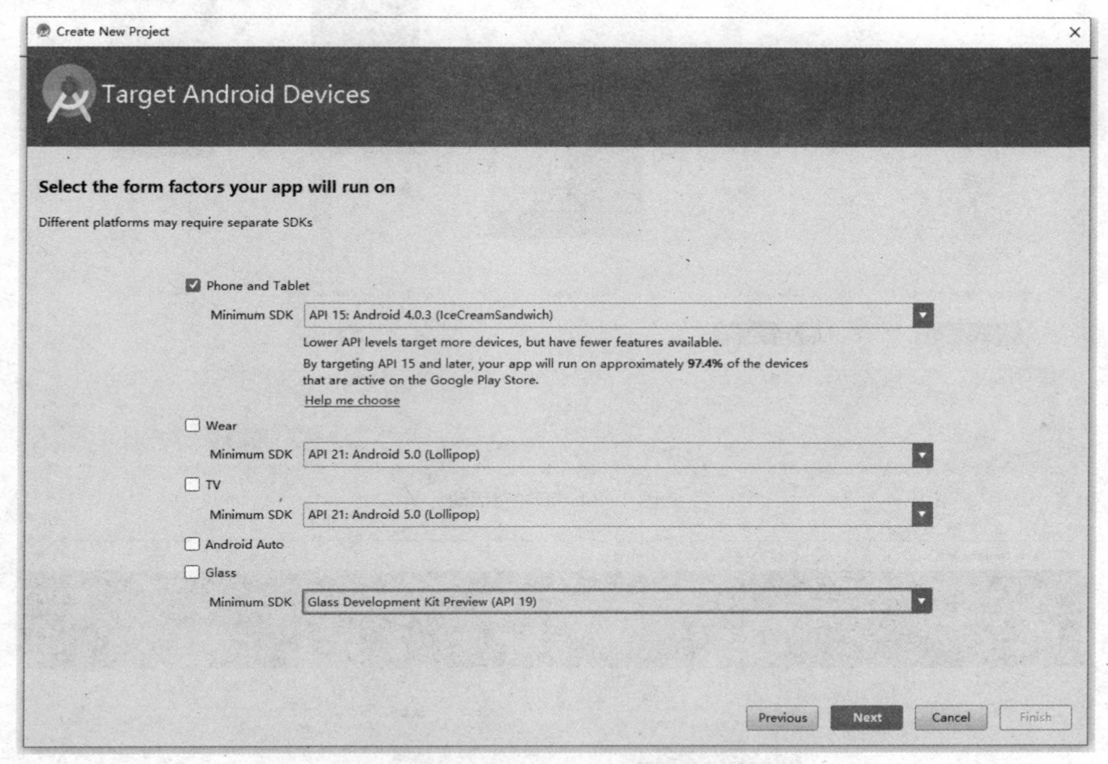

图 1-33　选择项目运行的目标设备

在指定存放项目文件的文件夹时，应注意文件夹路径不能包含非 ASCII 码字符，例如，不能将项目文件保存到路径含中文字符的文件夹中。

（3）在目标设备对话框中，可选择项目运行的目标设备。默认情况下勾选了 Phone and Tablet 复选框，表示项目可运行在 Android 手机和平板电脑上。在 Minimum SDK 列表中可选择项目兼容的最低的 Android SDK 版本。其他几个选项表示还可将项目运行在 Android 穿戴设备、Android 电视、Android 汽车和 Android 眼镜等多种设备上。目前，只需要勾选 Phone and Tablet 复选框，单击 Next 按钮，打开添加活动对话框，如图 1-34 所示。

（4）在添加活动对话框中显示了多种活动模板，选中 Empty Activity 来添加一个空活动，然后单击 Next 按钮，打开定义活动对话框，如图 1-35 所示。

（5）在定义活动对话框的 Activity Name 文本框中显示了默认的活动名称，可将其修改为其他名称。Layout Name 文本框中显示了活动使用的布局的默认名称，也可将其修改为其他名称。最后，单击 Finish 按钮，Android Studio 即可根据设置创建项目。

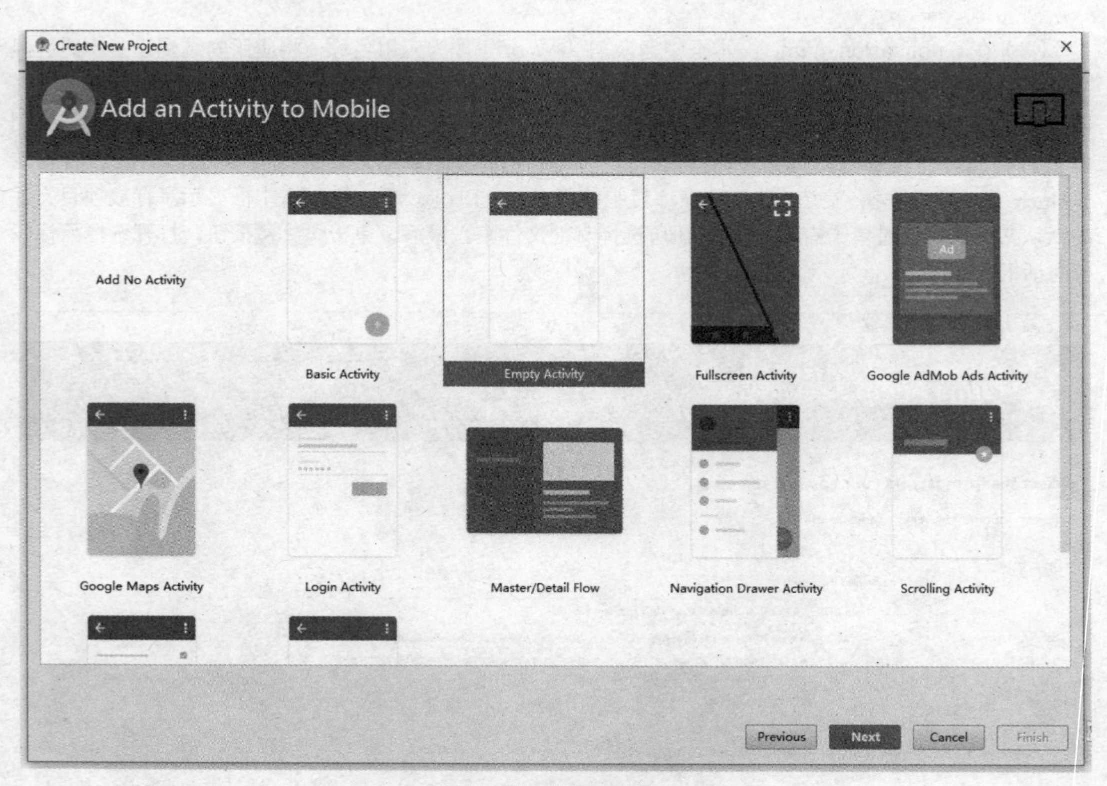

图 1-34　添加空活动

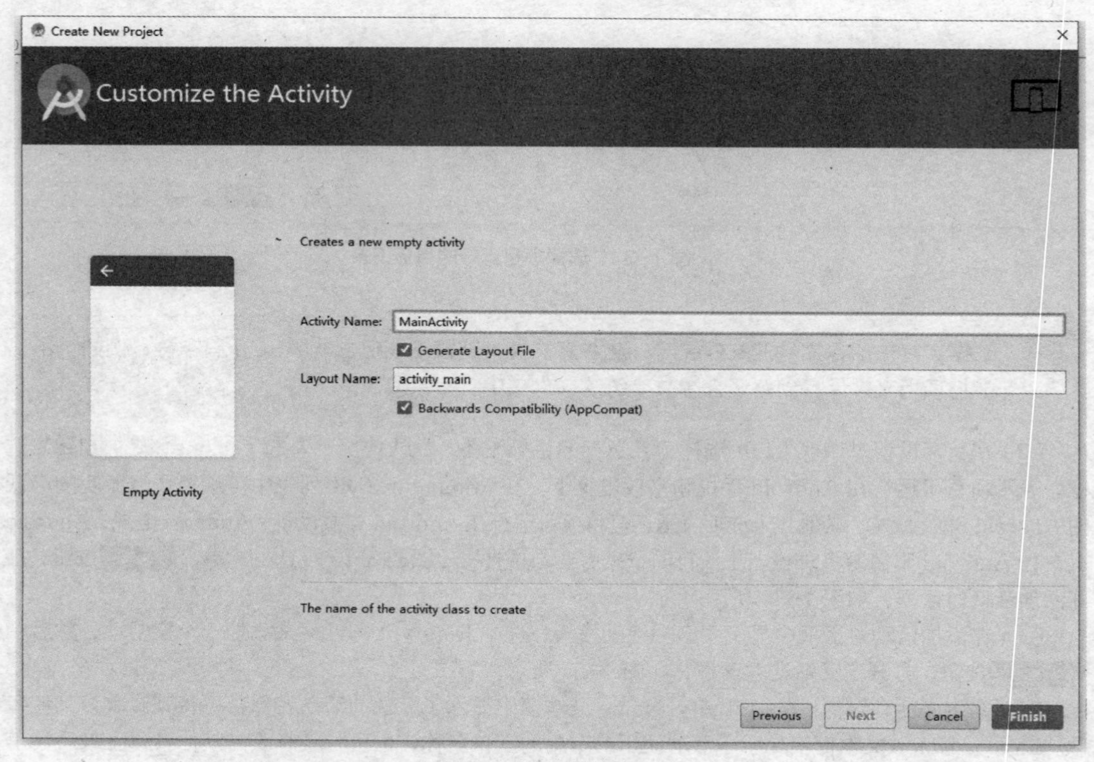

图 1-35　定义活动

Android Studio 使用 Gradle 脚本来构建项目，第一次创建项目时，可能需要从网络服务器下载相关组件，所以耗时略长。

在创建项目时，可能会遇到下面的错误提示。

Gradle sync failed: CreateProcess error=216, 该版本的 %1 与你运行的 Windows 版本不兼容。请查看计算机的系统信息，然后联系软件发布者。
Consult IDE log for more details (Help | Show Log)

这主要是因为 Android Studio 没有找到 JDK，而使用内置的 JDK 所引起。出现这种错误时，可在 Android Studio 中选择“File\Project Structure”命令，打开 Project Structure 对话框，如图 1-36 所示。

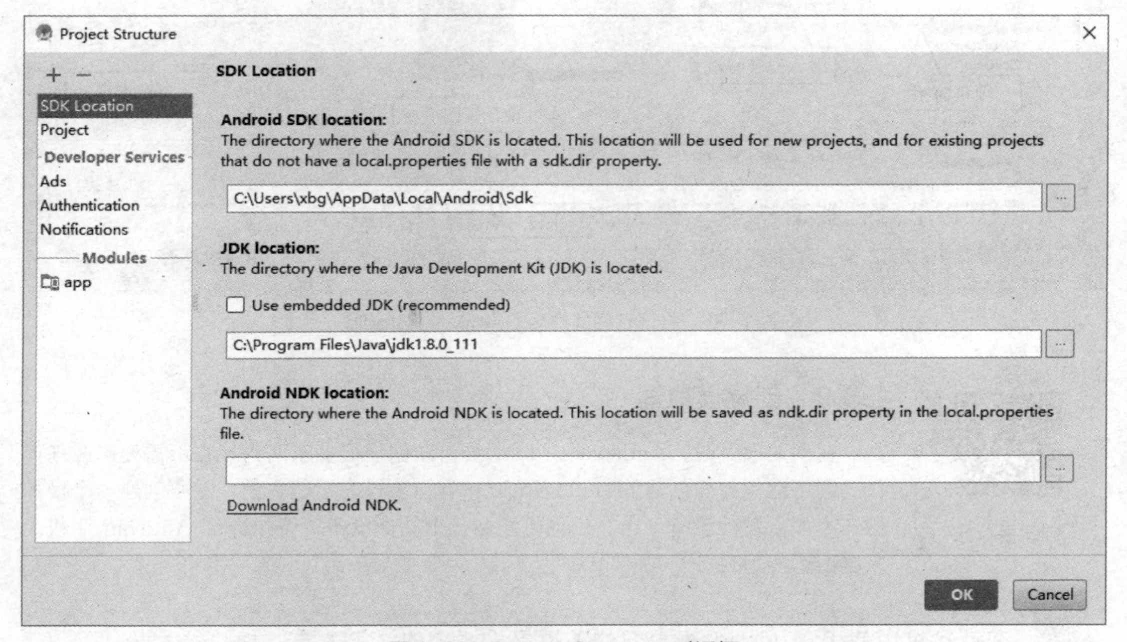

图 1-36　Project Structure 对话框

单击对话框左侧列表中的 SDK Location 选项，显示 SDK 位置设置。取消勾选 Use embedded JDK (recommended) 复选框，然后在下方的文本框中输入系统中 JDK 的安装路径，这里是 C:\Program Files\Java\jdk1.8.0_111。最后单击 OK 按钮保存设置。

图 1-37 显示了打开布局文件 activity_main.xml 后的 Android Studio 窗口，当前窗口主要显示了菜单栏、工具栏、项目窗格、组件列表窗格、设计视图窗格、属性窗格以及事件日志窗格等。

项目窗格可以用多种模式显示当前项目中的源代码文件、各种资源文件以及其他文件，双击文件可打开对应的编辑器。

组件列表窗格通常跟随布局文件的设计视图窗格一同显示，从组件列表窗格中将组件拖动到设计视图窗格即可为布局添加组件。

在设计视图窗格中以可视化的方式设计用户界面，实时显示界面的显示效果。在设计视图中选中某个组件时，属性窗格显示该组件的相关设置，并可修改组件的属性设置。

事件日志窗格显示了 Android Studio 执行各种操作的相关信息，例如，在应用运行出错时，事件日志窗格中就会显示相应的错误信息。

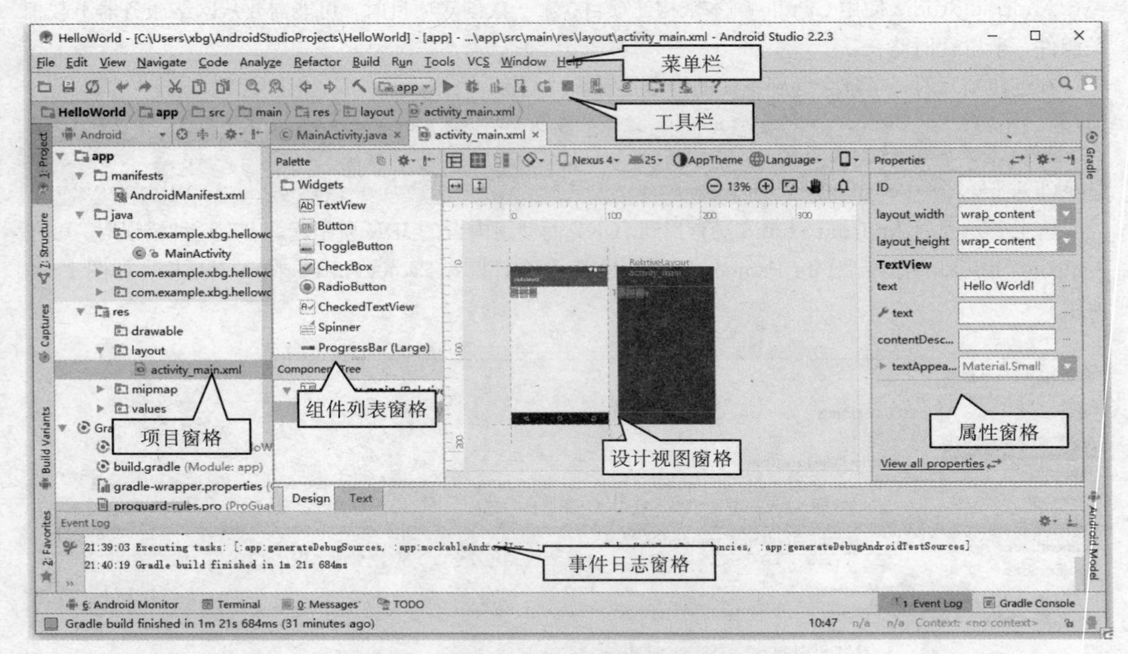

图 1-37　打开布局文件后的 Android Studio

创建模拟器

1.3.2　创建模拟器

在创建项目时，Android Studio 会自动创建很多东西，现在不需要修改任何代码即可运行前面创建的 HelloWorld 项目。不过在运行之前，需要创建一个模拟器作为项目运行设备。当然，也可连接一个物理设备（例如一台 Android 手机）来测试运行项目。

在计算机中使用 Android 模拟器的前提是机器中的 CPU 支持虚拟技术（Virtual Technology，VT），并在 BIOS 设置中启用。

在 Android Studio 中创建模拟器的操作步骤如下。

（1）选择"Tools\Android\AVD Manager"命令，打开 Android 虚拟设备管理器，如图 1-38 所示。由于还没有创建过虚拟设备，所以窗口中没显示任何设备。

（2）单击 + Create Virtual Device... 按钮，打开虚拟设备配置的选择尺寸对话框，如图 1-39 所示。

（3）在选择尺寸对话框中，首先需要在 Category（类型）列表中选择设备类型，默认选择 Phone。然后在右侧的列表框中选中已有的设备名称，使用对应的设备尺寸。最后单击 Next 按钮，打开选择系统镜像对话框，如图 1-40 所示。

（4）选择系统镜像对话框中列出了各种类型的系统镜像，显示了系统镜像的版本名称、API 级别、CPU 架构类型和支持的 Android 版本。如果系统镜像的版本名称后显示了 Download 超链接，则说明该镜像还没有下载，单击 Download 超链接即可打开 Android SDK 下载安装对话框，根据提示即可从网络服务器下载并安装系统镜像。由图 1-40 可见，应用于 Android 7.0 x86 的系统镜像 Nougat 已经安装到系统。选中该镜像，单击 Next 按钮，打开确认设置对话框，如图 1-41 所示。

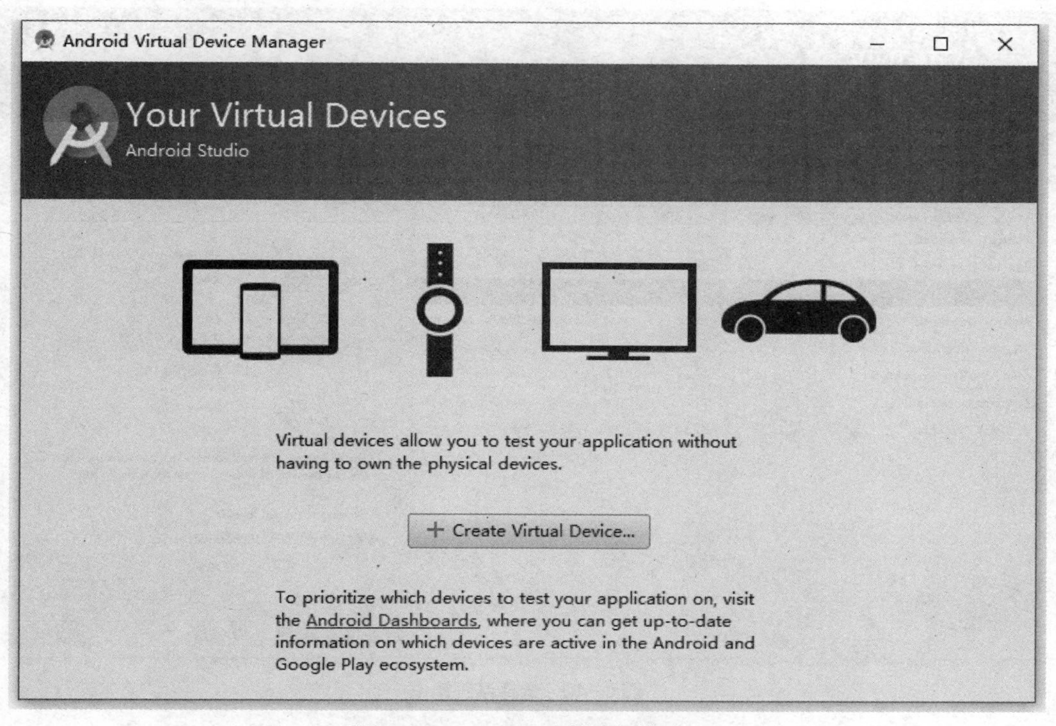

图 1-38　Android 虚拟设备管理器

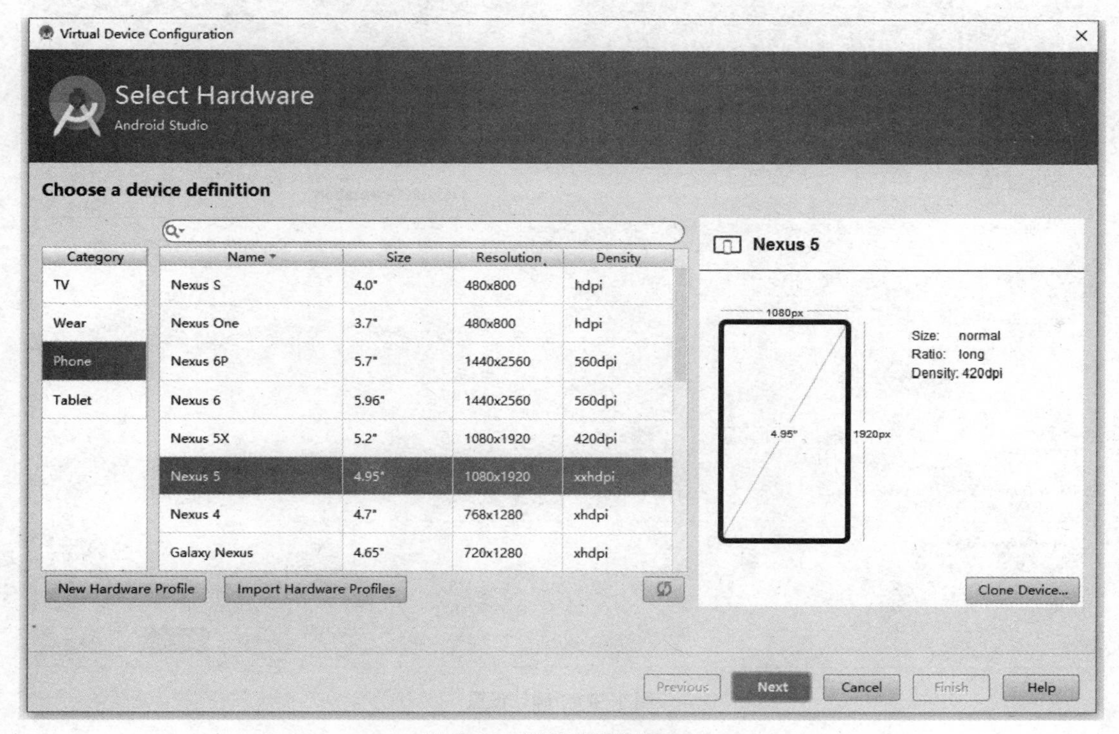

图 1-39　选择虚拟设备尺寸

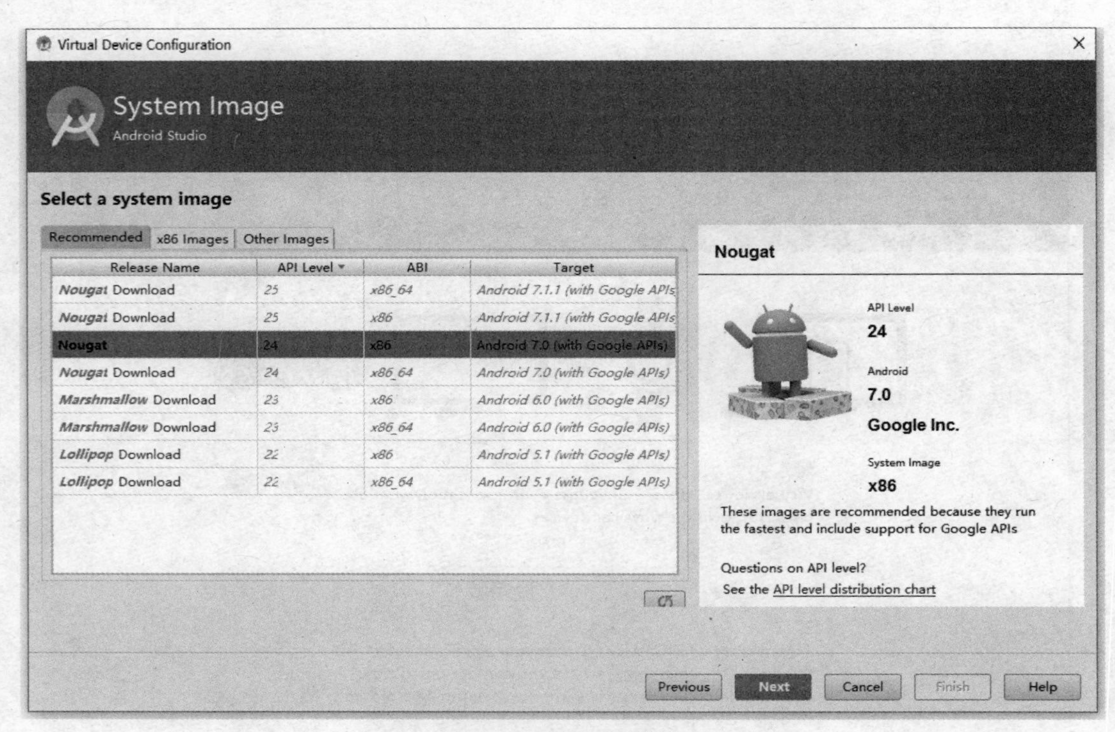

图 1-40　选择系统镜像

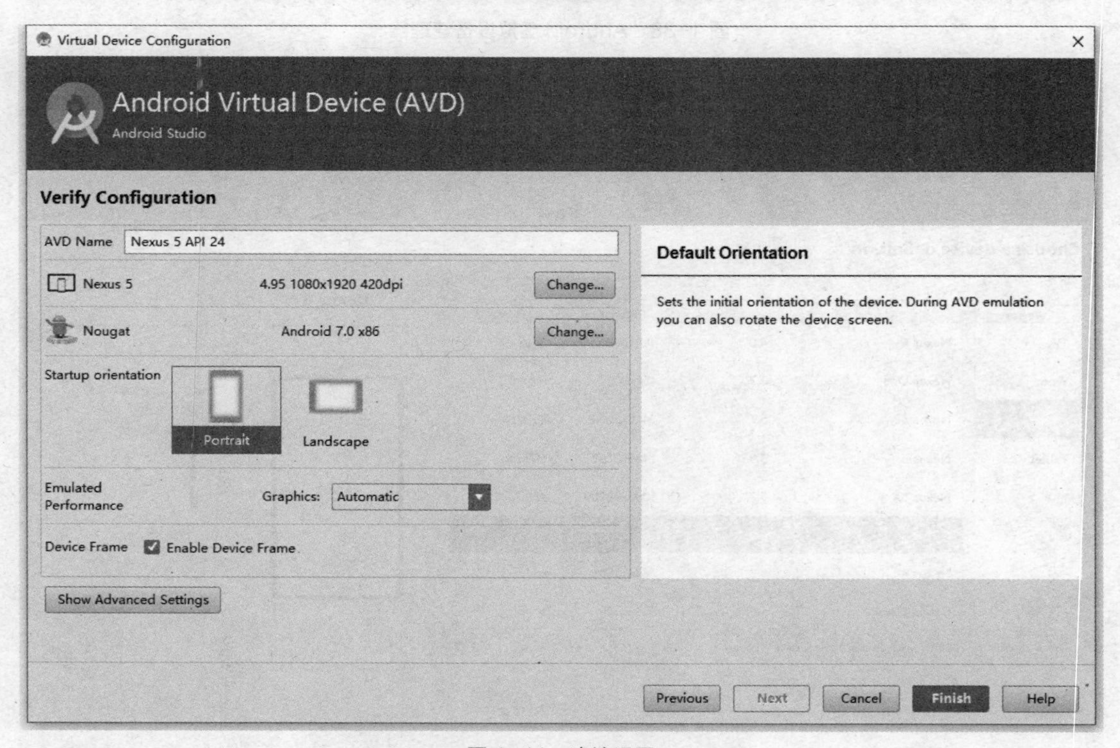

图 1-41　确认设置

（5）在确认设置对话框中，可在 AVD Name 文本框中修改模拟器名称。名称下面的两行分别显示了即将创建的模拟器尺寸和使用的系统镜像名称，单击 Change... 按钮，可重新选择模拟器尺寸和系统镜像。在 Startup orientation 框中，可选择设备打开时的方向。在 Emulated Performance 框中的 Graphics 下拉列表中，可选择使用硬件或软件来渲染图形，通常硬件渲染的速度更快。选项 Automatic 表示自动选择。默认情况下，Enable Device Frame 复选框为勾选状态，表示会为模拟器显示一个边框，使其更像真实设备。单击 Show Advanced Settings 按钮可显示高级选项，如图 1-42 所示，可设置摄像头、网络类型、存储器大小以及是否支持键盘输入等选项。最后，单击 Finish 按钮，Android Studio 即可根据设置创建模拟器。

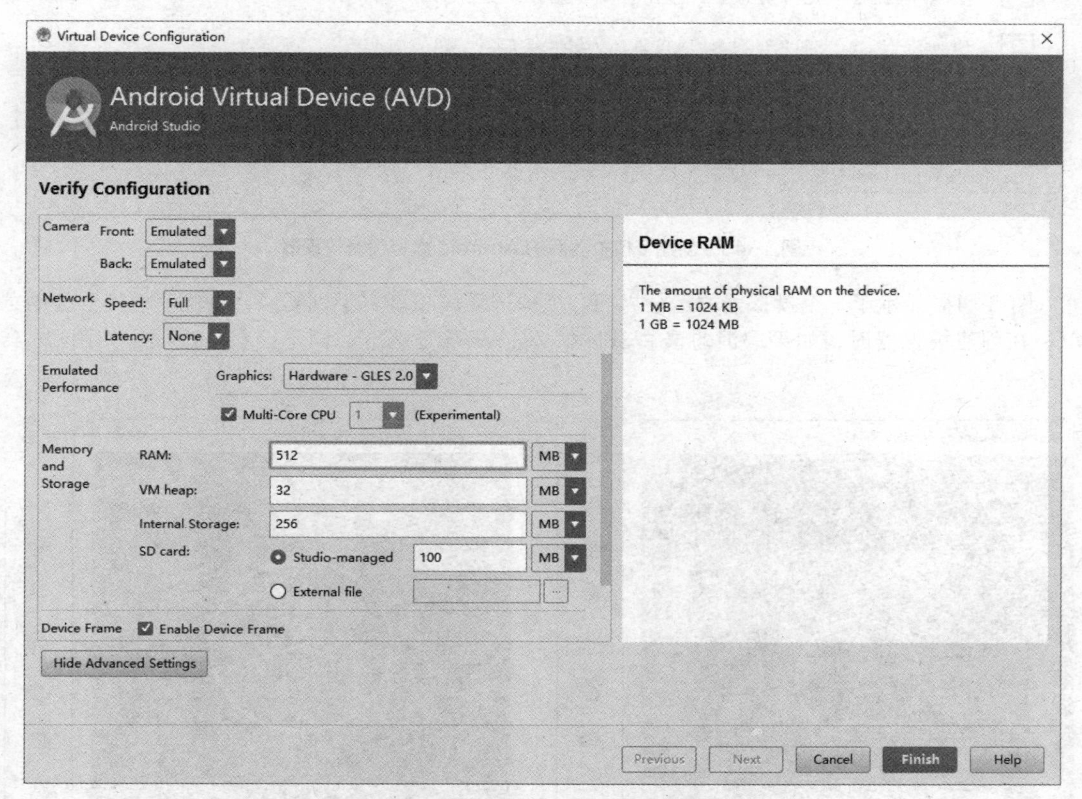

图 1-42　模拟器高级选项设置

使用模拟器的目的是为了测试 Android 应用程序，如果使用的计算机资源有限，可在创建模拟器时，将模拟器内存和内部存储器空间设置为较小的值，以便节约模拟器文件大小，加快创建和运行速度。

创建了模拟器后，Android 虚拟设备管理器窗口中会列出所有模拟器，如图 1-43 所示，显示模拟器的类型、名称、尺寸、API 级别、目标、CPU/ABI 类型及磁盘文件大小等信息。在每个设备的 Actions 列中可选择设备操作，具体如下。

- ▶：单击可启动模拟器，启动后可在 ▼ 菜单中选择停止。
- ✎：单击可打开图 1-42 所示的确认设置对话框，修改相关选项设置。
- ▼：单击可打开操作菜单，其中包含了 Duplicate（复制模拟器）、Wipe Data（擦除模拟器中已有的数据）、Show on Disk（显示模拟器在磁盘中的文件）、View Details（显示模拟器详细信息）、Delete（删除模拟器）和 Stop（停止已启动的模拟器）等命令。

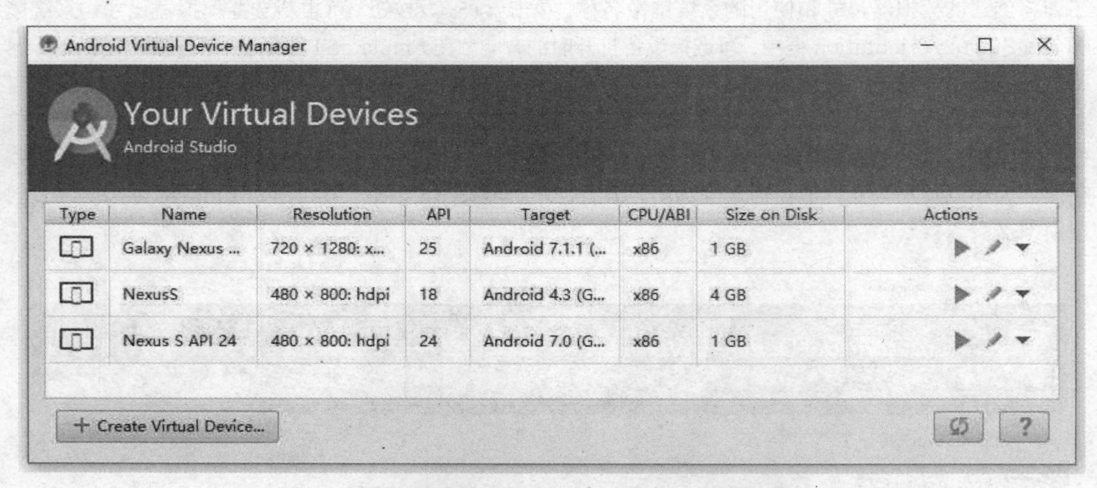

图 1-43　创建了模拟器后的 Android 虚拟设备管理器

图 1-44 显示了一个模拟器的运行界面。模拟器界面和真实手机类似，右侧为模拟器控制菜单。在创建模拟器时，如果取消勾选 Enable Device Frame 复选框，则不会显示边框，如图 1-45 所示。

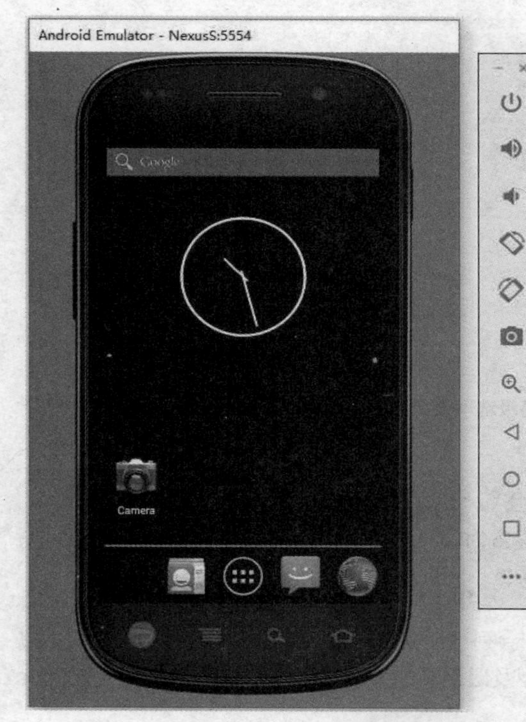

图 1-44　运行中的模拟器

图 1-45　无边框的模拟器

如果使用的计算机 CPU 不支持虚拟技术，可以利用 Android Studio 提供的其他解决办法。图 1-46 显示了 CPU 不支持虚拟技术的情况下的 Android 虚拟设备管理器。

可以看到，已创建的模拟器列表上方提示了系统 CPU 不支持的虚拟技术，单击右侧的 Troubleshoot 链接，可打开对话框供用户查看 Android Studio 推荐的解决办法，如图 1-47 所示。

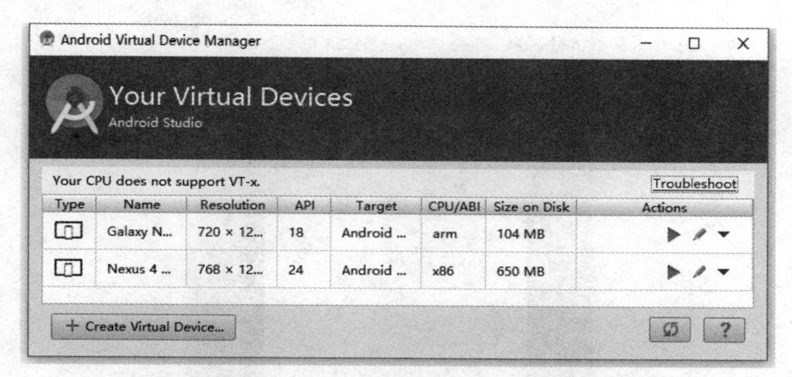

图 1-46　CPU 不支持虚拟技术时的 Android 虚拟设备管理器

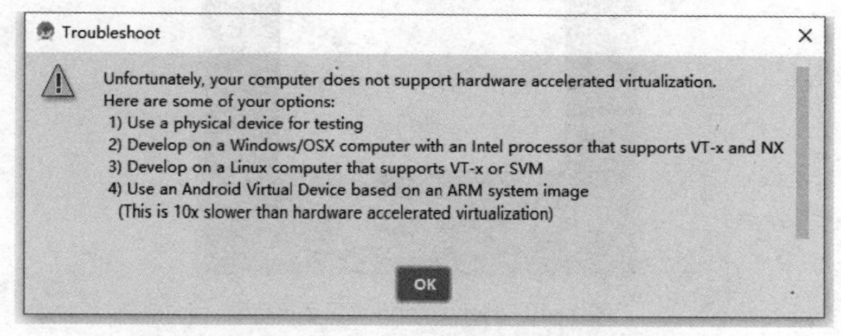

图 1-47　CPU 不支持虚拟技术时的解决办法

选择 1）表示连接物理设备，选择 2）、3）表示换一台 CPU 支持虚拟技术的计算机，选择 4）表示用 ARM 系统镜像来创建虚拟器（不过这种模拟器运行很慢）。

事实上，找一台 Android 手机或者平板电脑来调试 Android 程序是不错的选择。毕竟，除了真实的优点外，物理设备的运行速度是模拟器无法比拟的。开启 Android 设备的 USB 调试功能，通过 USB 连接到计算机后，便可用其调试 Android 应用了。

1.3.3　运行项目

单击 Android Studio 工具栏中的▶按钮，或选择“Run\Run App”命令，或者按【Shift+F10】组合键，即可运行应用。运行时，首先会打开图 1-48 所示的选择部署目标对话框。

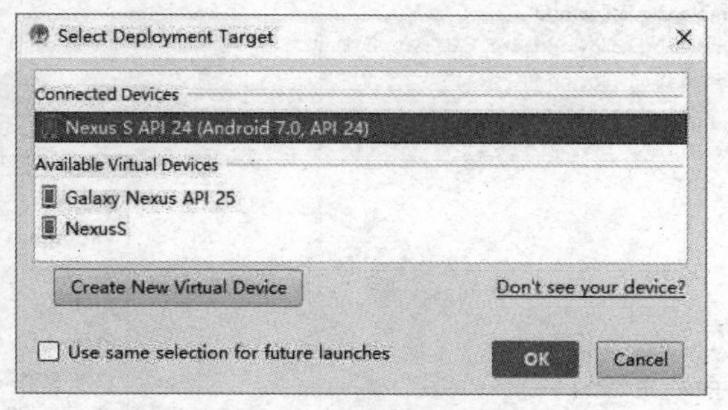

图 1-48　选择部署目标

部署目标就是用于测试运行应用的物理设备或模拟器，对话框中会列出当前连接的物理设备和模拟器。选中设备后，单击 OK 按钮，Android Studio 会将应用部署到设备中，并启动应用。图 1-49 显示了在 1.3.1 节中创建的 HelloWorld 项目的运行结果。

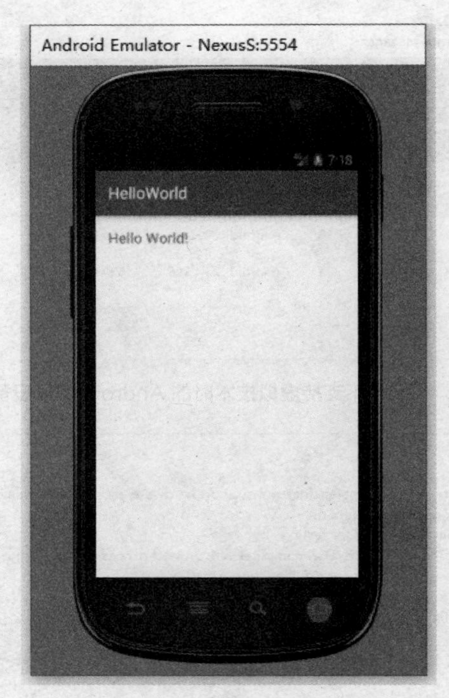

图 1-49 HelloWorld 项目运行结果

1.3.4 了解 Android 项目组成

1. 项目窗格

新建或打开一个项目后，单击 Android Studio 左侧边栏中的 Project 按钮，可打开或关闭项目窗格。图 1-50 显示了以 Android 模式显示的项目窗格。标题栏中的 Android 表示以 Android 模式显示项目结构。单击标题栏，可打开模式切换菜单，如图 1-51 所示。

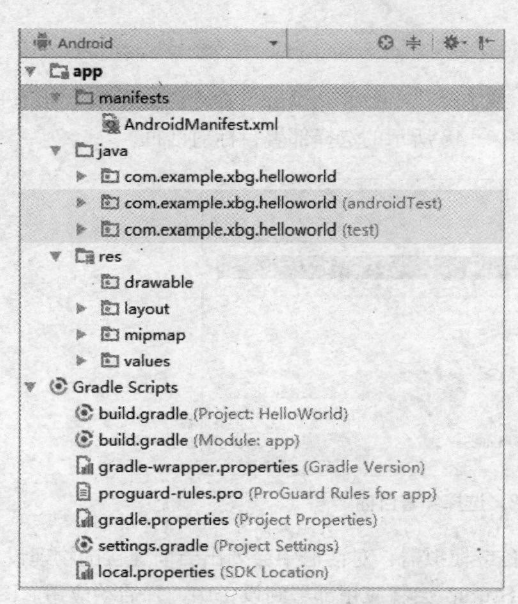

图 1-50 以 Android 模式显示的项目窗格

图 1-51 模式切换菜单

默认情况下，Android Studio 使用 Android 模式显示项目结构，Android 模式也最适合开发。在创建项目时，Android Studio 会自动生成很多文件，Android 模式下显示的文件都是项目中允许用户修改的常用文件，其他一些自动生成、不允许修改的文件都隐藏起来。

2. 顶层目录和文件

在 Android 模式下，开发人员可以快速找到需要的文件。Android 模式不能反映实际的项目文件目录结构。在项目窗格显示模式切换菜单中选择 Project，可以项目模式显示项目结构。图 1-52 所示为项目模式下 HelloWorld 项目对应的实际磁盘文件目录结构。

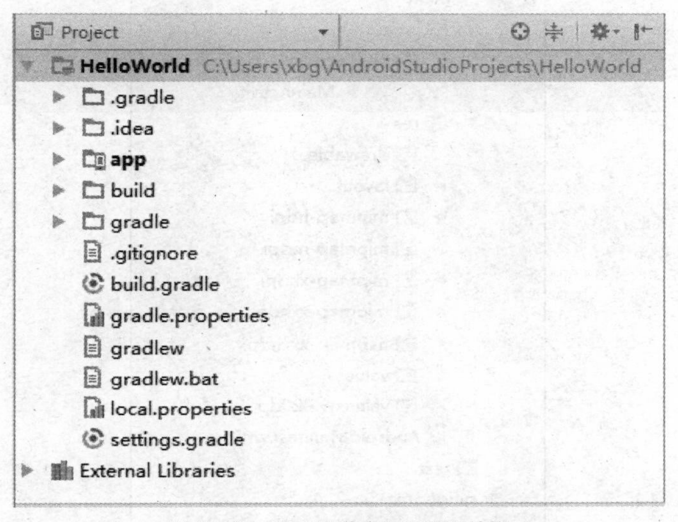

图 1-52　以项目模式显示项目结构

项目模式下，HelloWorld 项目顶层的各个目录和文件的作用如下。

- .gradle 目录：Android Studio 使用 Gradle 来构建项目，项目使用的 Gradle 程序的相关文件放在.gradle 目录中。
- .idea 目录：存放 Android Studio 的相关配置文件。
- app 目录：一个 Android Studio 项目可以包含多个模块。创建项目时，第一个模块会自动命名为 app。app 目录存放该模块本身使用的相关文件，包括源代码、资源及其他相关文件。
- build 目录：存放编译项目时自动生成的项目全局文件。项目各个模块的编译文件保存在模块的 build 子目录中。
- gradle 目录：存放 Gradle 脚本和相关配置文件。
- .gitignore 文件：文件中的目录和文件将排除在项目全局的版本控制之外。
- build.gradle 文件：项目全局的 Gradle 脚本。
- gradle.properties 文件：项目全局的 Gradle 配置文件。
- gradlew 文件：在 Mac、Linux 等系统命令行执行的 Gradle 脚本。
- gradlew.bat 文件：在 Windows 系统命令行执行的 Gradle 脚本。
- local.properties 文件：Gradle 使用的 Android SDK 路径配置文件，Android Studio 自动生成的，不允许用户修改。
- settings.gradle 文件：设置项目中包含的模块名称，默认情况下，项目只有一个 app 模块，所以文件中只包含 app。通常，项目中添加的所有模块都会自动包含在文件中。

3. app 目录和文件

在项目顶层的目录和文件中，除了 app 目录外，其他目录和文件都是 Android Studio 自动生成的，一般情况下也不需要手动修改，项目设计开发主要集中在 app 目录中。app 目录结构如图 1-53 所示。

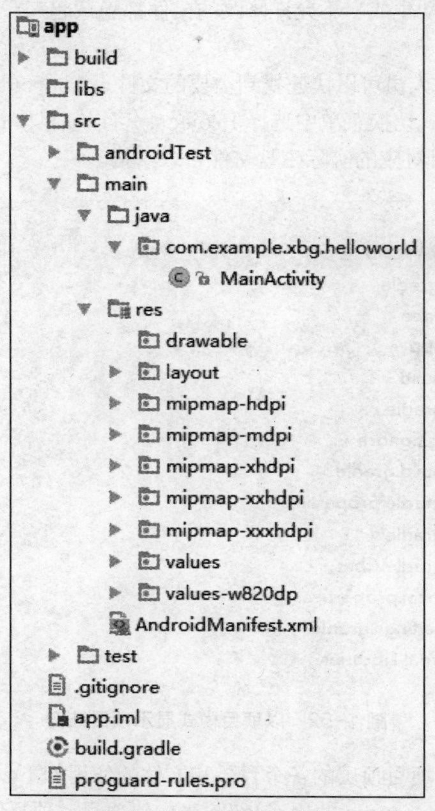

图 1-53　app 目录结构

app 的各个目录和文件作用如下。

- build 目录：存放模块在编译时生成的文件。
- libs 目录：存放项目中使用的第三方 Java 库文件。
- src\androidTest 目录：存放执行 Instrumented 测试用例的文件。
- src\main 目录：存放源代码相关文件。java 目录中的内容为模块 Java 源代码文件。其中，com.example.xbg.helloworld 为项目的包名称，MainActivity 为活动的源代码文件。main 目录中的 AndroidManifest.xml 为模块清单文件。
- res\drawable 目录：存放图片资源文件。
- res\layout 目录：存放布局文件。
- res\mipmap-hdpi、res\mipmap-mdpi、res\mipmap-xhdpi、res\mipmap-xxhdpi、res\mipmap-xxxhdpi 等目录：存放各种分辨率下的项目的图标文件。
- res\values 目录：存放颜色、尺寸、字符串和样式等资源文件。
- res\values-w820dp 目录：存放屏幕尺寸不小于 820dp 时的自定义尺寸资源文件。
- src\test 目录：存放执行 UNIT 测试用例的文件。
- .gitignore 文件：文件中的目录和文件将排除在模块的版本控制之外。
- app.iml：IntelliJ IDEA 项目自动生成的标识文件。
- build.gradle 文件：模块的 Gradle 脚本。

- proguard-rules.pro 文件：项目文件的混淆规则文件。在生成项目安装包时使用混淆规则，可使破解者难以阅读程序代码。

4. 清单文件 AndroidManifest.xml

每个 Android 应用都有一个清单文件，该文件必须放在应用项目的根目录中。每个项目中的 src\main 目录为代码的根目录。在 Android 应用中启动组件时，都会先从清单文件中确认组件是否存在。清单文件的主要作用如下所述。

- 声明应用中的所有组件。应用中的活动、服务、广播接收器和内容提供程序等都必须在清单文件中声明。
- 声明用户权限，例如读取用户联系人权限、网络访问权限等。
- 声明应用最低兼容的 API 级别。
- 声明应用可访问的软硬件功能，例如摄像头、蓝牙或多点触摸屏等。
- 声明应用需链接的 API 库，例如 Google 地图库。

在 1.3.1 节中创建的 HelloWorld 项目的 AndroidManifest.xml 文件代码如下。

```
<?xml version="1.0" encoding="utf-8"?>
<manifest xmlns:android="http://schemas.android.com/apk/res/android"
    package="com.example.xbg.helloworld" >
    <application
        android:allowBackup="true"
        android:icon="@mipmap/ic_launcher"
        android:label="@string/app_name"
        android:supportsRtl="true"
        android:theme="@style/AppTheme" >
        <activity android:name=".MainActivity" >
            <intent-filter>
                <action android:name="android.intent.action.MAIN" />
                <category android:name="android.intent.category.LAUNCHER" />
            </intent-filter>
        </activity>
    </application>
</manifest>
```

其中，application 元素声明了应用程序配置，android:allowBackup 设置应用是否支持备份和恢复功能，android:icon 设置应用的图标，android:label 设置应用的标题，android:supportsRtl 设置应用是否支持从右到左布局（RTL：right-to-left），android:theme 设置应用的主题样式。在设置各个属性时，可以使用直接的值，例如 true；也可使用资源文件中定义的资源名称，例如@string/app_name。@string/app_name 表示字符串资源文件 string.xml 中名为 app_name 的字符串。

<activity>用于元素声明项目中的一个活动，android:name 属性指定活动名称。<intent-filter>元素设置活动使用的 intent 对象。<action android:name="android.intent.action.MAIN" />和<category android:name="android.intent.category.LAUNCHER" />表示活动为主活动，在设备中单击应用程序图标时，首先运行该活动。

<activity>的 android:name 属性指定活动的类名称为 MainActivity，名称之前的句点符号表示类名称的限定符为当前包名，完整的名称为 com.example.xbg.helloworld.MainActivity。

5. 活动源代码文件 MainActivity.java

MainActivity.java 为活动的 Java 类实现代码文件，其代码如下。

```
package com.example.xbg.helloworld;
import android.support.v7.app.AppCompatActivity;
import android.os.Bundle;
public class MainActivity extends AppCompatActivity {
    @Override
    protected void onCreate(Bundle savedInstanceState) {
        super.onCreate(savedInstanceState);
        setContentView(R.layout.activity_main);
    }
}
```

其中，package 语句指定了包名称，import 语句指定了要导入的类。活动 MainActivity 继承了 AppCompatActivity，AppCompatActivity 是所有活动的基类 Activity 的子类。onCreate()方法在创建活动时执行，super.onCreate()方法表示调用父类的 onCreate()方法创建活动实例对象。在后继的很多知识点学习过程中，都会在 onCreate()方法中编写代码来完成相关操作。

setContentView()方法设置活动加载的布局文件，R.layout.activity_main 表示布局文件为 res\layout 中的 activity_main.xml。布局文件是活动启动后展示给用户的图形界面，从这里也看出，Android 程序的设计强调将功能逻辑和界面设计分离。

6. 布局文件 activity_main.xml

布局文件是一个 XML 文件，它为活动定义用户界面。activity_main.xml 代码如下。

```
<?xml version="1.0" encoding="utf-8"?>
<RelativeLayout
    xmlns:android="http://schemas.android.com/apk/res/android"
    xmlns:tools="http://schemas.android.com/tools"
    android:id="@+id/activity_main"
    android:layout_width="match_parent"
    android:layout_height="match_parent"
    android:paddingLeft="@dimen/activity_horizontal_margin"
    android:paddingRight="@dimen/activity_horizontal_margin"
    android:paddingTop="@dimen/activity_vertical_margin"
    android:paddingBottom="@dimen/activity_vertical_margin"
    tools:context="com.example.xbg.helloworld.MainActivity">

    <TextView
        android:layout_width="wrap_content"
        android:layout_height="wrap_content"
        android:text="Hello World!" />
</RelativeLayout>
```

<RelativeLayout>为布局文件的根元素，对于 RelativeLayout，这里也表示使用相对布局。android:id 属性设置了活动的 ID，在项目的其他文件中都可以通过该 ID 来引用活动。android:layout_width 设置了布局的宽度，match_parent 在这里表示布局宽度与运行项目的设备宽度一致。android:layout_height 设置布局的高度，match_parent 在这里表示布局高度与运行项目的设备高度一致。android:paddingLeft 设置布局的左边距，android:paddingRight 设置布局的右边距，android:

paddingTop 设置布局的顶部边距，android:paddingBottom 设置布局的底部边距，tools:context 设置上下文范围。

<TextView>元素为布局添加一个文本视图控件。android:layout_width 属性设置控件宽度，wrap_content 表示控件宽度刚好容纳内容。android:layout_height 属性设置控件高度，wrap_content 表示控件高度刚好容纳内容。android:text 设置控件显示的字符串，"Hello World!"为显示在文本视图控件中的字符串。如果需要显示其他字符串，修改 android:text 设置即可。

通过前面的分析可以了解到 Android 应用程序执行的基本过程：在启动 Android 应用时，应用首先启动一个主活动，主活动加载一个布局，布局用各种控件将字符串等各种内容呈现给用户。

1.4 Android 编程小工具——日志

在程序中，可使用 System.out、System.err 以及 Log 对象等多种方法输出调试信息，供开发人员查看程序运行状态。

1.4.1 使用日志 API 输出调试信息

可使用下面的多种方法在程序中输出调试信息，这些信息统称为日志，具有不同的级别。

使用日志 API

- System.out.println()：输出的日志级别为 Info，即普通信息。
- System.err.println()：输出的日志级别为 Warn，即警告信息。
- Log.v()：输出的日志级别为 Verbose，即冗余信息。
- Log.d()：输出的日志级别为 Debug，即调试信息。
- Log.i()：输出的日志级别为 Info，即普通信息。
- Log.w()：输出的日志级别为 Warn，即警告信息。
- Log.e()：输出的日志级别为 Erro，即错误信息。

Log 对象的各个方法的第一个参数为日志 Tag，第二个参数为日志内容。在查看日志时，可使用 Tag 来分类显示日志。

修改前文 HelloWorld 项目中的 MainActivity.java，在 OnCreate()方法中添加上述方法来输出相应的日志信息，代码如下。

```
package com.example.xbg.helloworld;
import android.support.v7.app.AppCompatActivity;
import android.os.Bundle;
import android.util.Log;
public class MainActivity extends AppCompatActivity {
    @Override
    protected void onCreate(Bundle savedInstanceState) {
        super.onCreate(savedInstanceState);
        setContentView(R.layout.activity_main);
        System.out.println("System.out输出");
        System.err.println("System.err输出");
        Log.v("MainActivity","Log.v输出");
        Log.d("MainActivity","Log.d输出");
        Log.i("MainActivity","Log.i输出");
```

```
        Log.w("MainActivity","Log.w输出");
        Log.e("MainActivity","Log.e输出");
    }
}
```

运行项目后，在 Logcat 窗口中可看到输出的多条日志信息，如图 1-54 所示。

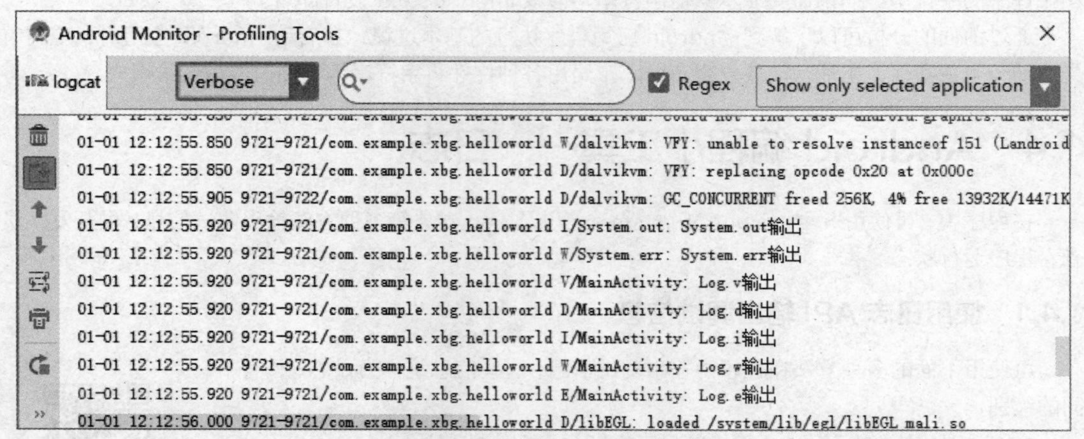

图 1-54　输出日志信息

1.4.2　日志分类与日志过滤器

Android 应用程序日志可分 5 个级别，从低到高依次为 Verbose、Debug、Info、Warn 和 Erro。

Logcat 窗口默认显示不低于 Verbose 级别的日志，即显示全部日志信息。在 Logcat 窗口的日志级别列表 Verbose 中，选择其他级别，即可过滤掉低于该级别的信息。例如，图 1-55 中即显示了不低于 Warn 级别的所有日志信息。

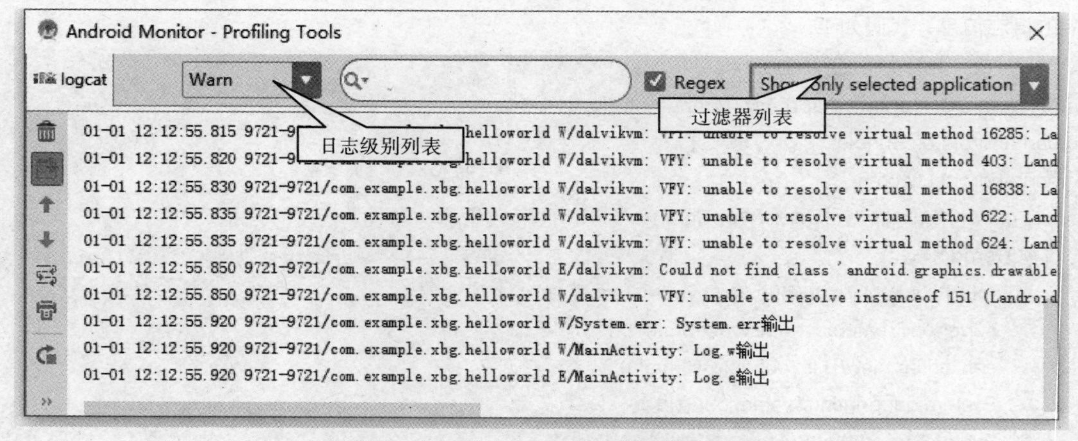

图 1-55　显示不低于 Warn 级别的所有日志信息

除了通过日志级别分类显示日志信息外，还可使用过滤器筛选日志信息。Logcat 窗格右上角为过滤器列表，默认过滤器为 Show only selected application，即仅显示当前选中应用程序的信息。如果在过滤器列表中选中 Edit Filter Configuration 选项，可打开创建过滤器对话框，如图 1-56 所示。

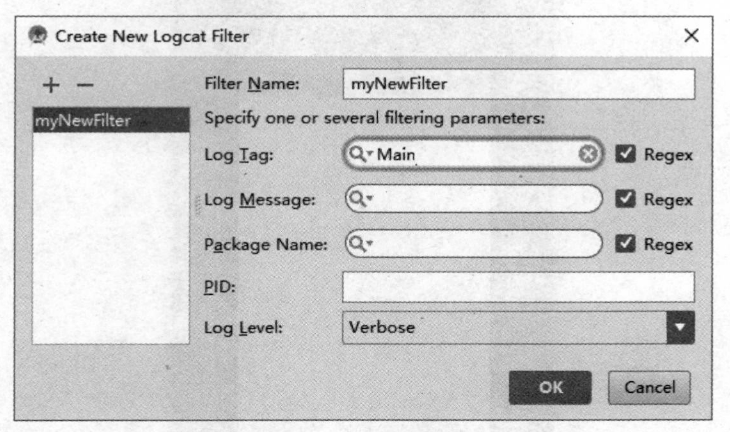

图 1-56 创建过滤器

创建新过滤器时，首先在 Filter Name 文本框中输入过滤器名称，例如 myNewFilter。在 Log Tag 框中可输入日志 Tag 的筛选字符串，例如，Main 表示只显示 Tag 中包含 Main 的日志信息。在 Log Message 框中可输入日志内容的筛选字符串，在 Package Name 框中可输入包名称的筛选字符串。在 PID 框中可输入应用程序的进程 ID。在 Log Level 列表中，可选择筛选器应用的最低日志级别，低于该级别的信息不显示。最后，单击 OK 按钮即可应用新建的过滤器。图 1-57 显示了应用图 1-56 中创建的 myNewFilter 后，Logcat 窗格中的信息。

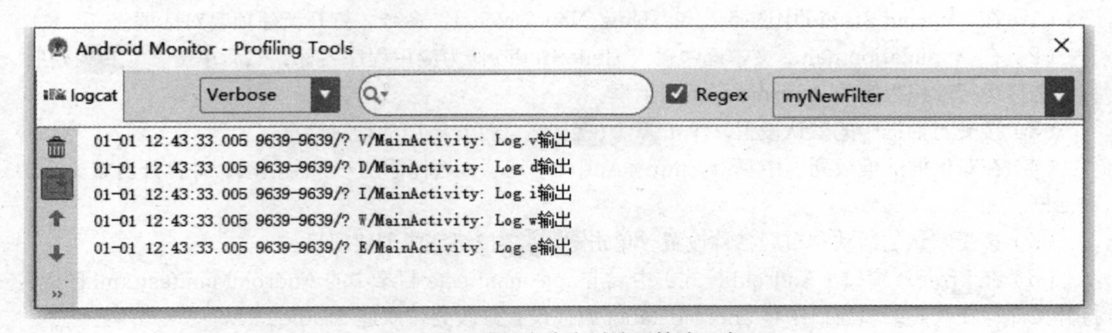

图 1-57 应用自定义过滤器筛选日志

在创建过滤器对话框中，左侧的列表会显示已有的过滤器名称，选中过滤器，即可在右侧显示过滤器的对应设置项，可修改现有的过滤器。单击过滤器列表框上方的➕按钮，可添加新的过滤器。单击➖按钮，可删除当前选中的过滤器。

新建的过滤器名称会出现在 Logcat 窗格的过滤器列表中，在列表中选中即可使用该过滤器来筛选日志信息。

1.5 编程实践：你好，Android Studio!

本节综合应用本章所学知识，在 Android Studio 中创建一个项目，在手机中运行时，屏幕上显示“你好，Android Studio!”，如图 1-58 所示。

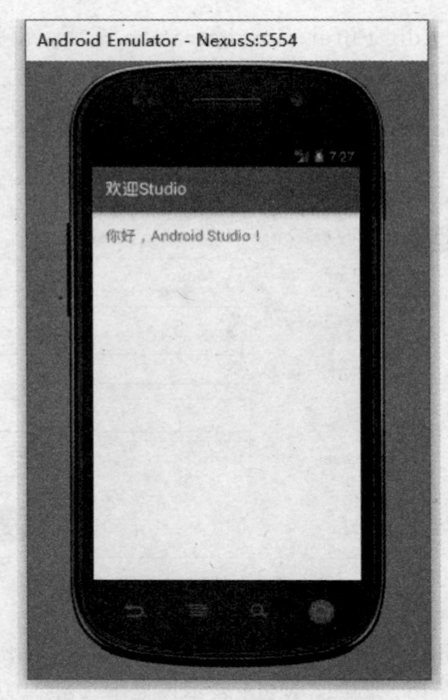

图 1-58　编程实践实例运行效果

1.5.1　创建 HelloStudio 应用

具体操作步骤如下。

（1）在 Android Studio 中选择“File\New\New Project”命令，打开新建项目对话框。

（2）在 Application name 文本框中输入 HelloStudio 作为应用程序名称，然后单击 Next 按钮，打开选择运行应用程序的 Android 设备对话框。

（3）接受对话框中的默认设置，单击 Next 按钮，打开添加活动对话框。

（4）在对话框的模板列表中双击 Empty Activity 选项，选择添加一个空活动，并打开自定义活动对话框。

（5）接受默认的活动和布局名称设置，单击 Finish 按钮完成创建项目。

（6）在 Project 窗口（Android 模式）中双击 app\manifests 目录中的 AndroidManifest.xml 图标，打开文件，在代码编辑窗口中将 android:label 属性设置修改为“欢迎 Studio”。

（7）在 Project 窗口（Android 模式）中双击 app\res\values 目录中的 strings.xml 图标，打开文件，在代码编辑窗口中添加一条字符串定义，代码如下（加粗部分为添加代码）。

```
<resources>
    <string name="app_name">HelloStudio</string>
    <string name="showmsg">你好，Android Studio！</string>
</resources>
```

（8）在 Project 窗口（Android 模式）中双击 app\res\layout 目录中的 activity_main.xml 图标，打开文件，在代码编辑窗口中将<TextView>元素的 android:text 属性设置为@string/showmsg，即显示字符串资源文件中定义的 showmsg 字符串。

（9）按【Shift+F10】组合键运行项目，打开选择部署目标对话框。

（10）在对话框中双击用于运行项目的设备，Android Studio 即可将项目部署到设备并运行，运行结果如图 1-58 所示。

1.5.2 打包发布 APK 安装包

在使用真机设备进行调试时，Android Studio 会自动完成一系列操作，包括将应用程序代码打包成 APK 文件、将 APK 文件发送到设备、Android 系统识别 APK 文件执行安装操作、运行应用等步骤。

Android 系统只允许经过签名的 APK 文件进行安装。在调试时，Android Studio 使用了一个默认的签名文件 debug.keystore，该文件默认在系统当前用户文件夹.android 中（例如 C:\Users\Administrator\.android）。

用户可创建自己的签名文件来生成 APK，具体操作步骤如下。

（1）在 Android Studio 中选择“Build\Generate Singed APK”命令，打开生成签名 APK 对话框，如图 1-59 所示。

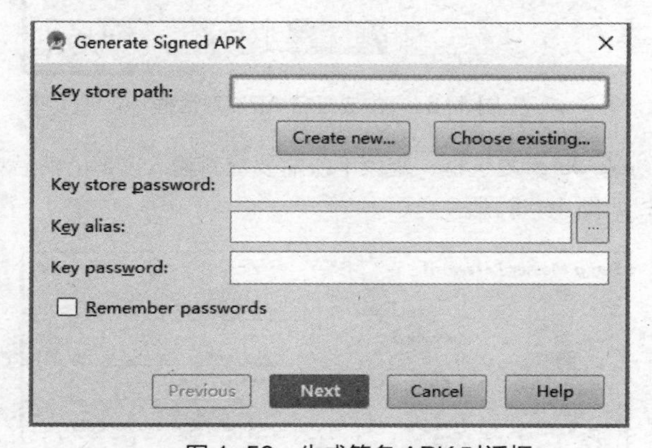

图 1-59　生成签名 APK 对话框

（2）目前还没有用于签名的 Key Store 文件，所以需创建一个。单击 Create new... 按钮，打开新建 Key Store 文件对话框，如图 1-60 所示。

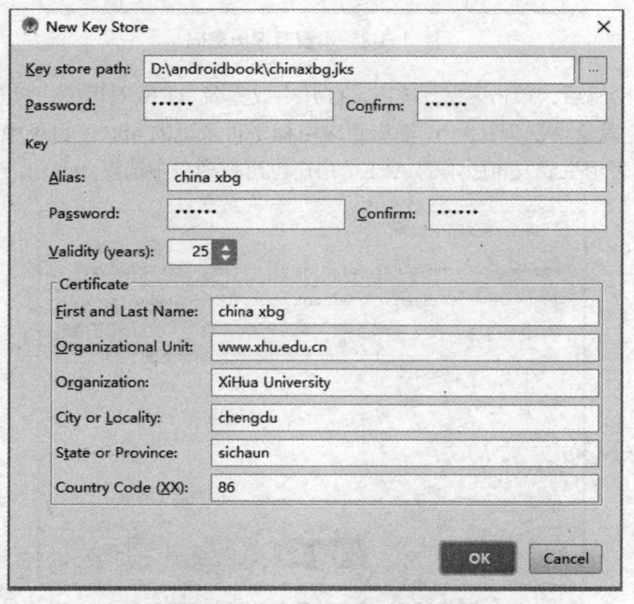

图 1-60　添加 Key Store 文件信息

（3）按要求填写 Key Store 文件信息，最后单击 OK 按钮生成 Key Store 文件。Android Studio 会自动生成签名 APK 对话框中使用的新建 Key Store 文件，如图 1-61 所示。

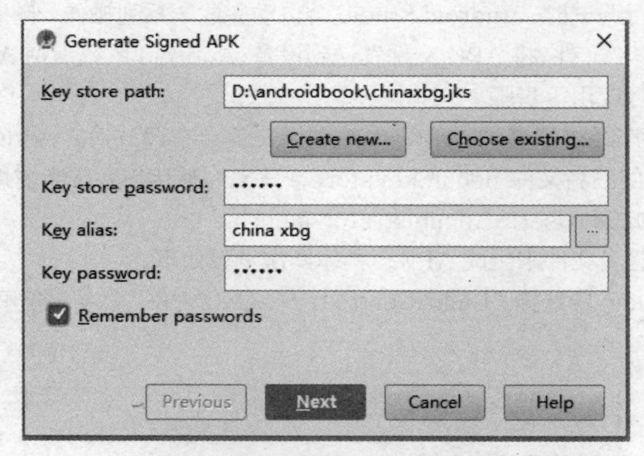

图 1-61　生成签名 APK 对话框

（4）勾选 Remember passwords 复选框，避免下次需要重新输入密码。然后单击 Next 按钮，打开设置管理员密码对话框，如图 1-62 所示。

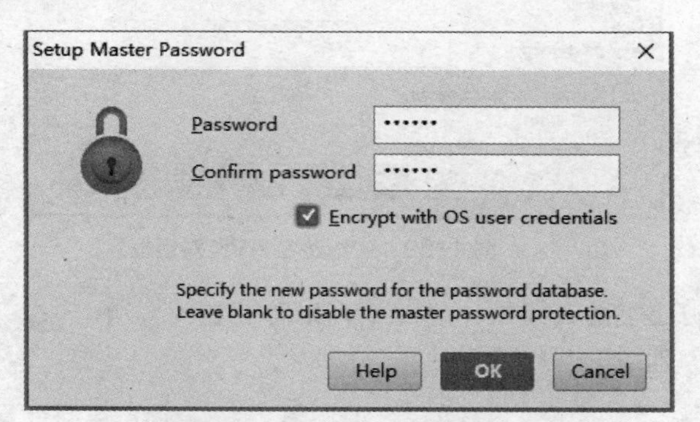

图 1-62　设置管理员密码

（5）输入管理员密码后，单击 OK 按钮，打开生成签名 APK 对话框，设置 APK 文件的输出路径，如图 1-63 所示。默认情况下，APK 输出到应用程序目录下的 app 子目录中。在 Build Type 下拉列表中选择 release，表示生成发布版本的 APK。用户也可在列表中选择 debug，生成调试版本的 APK。

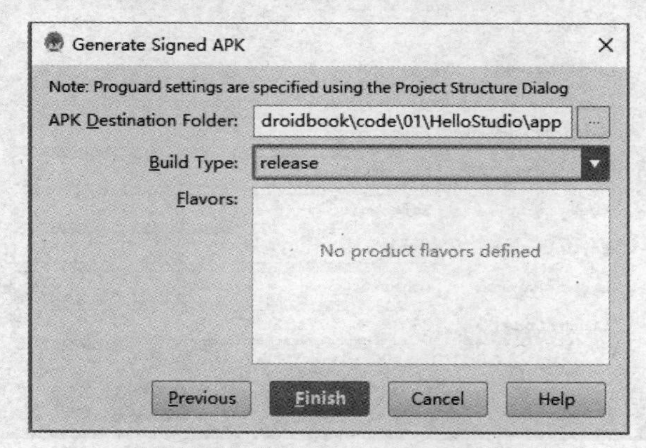

图 1-63　设置 APK 文件输出路径

（6）单击 Finish 按钮，Android Studio 即可生成签名的 APK。APK 文件创建完成后，Android Studio 会在右上角显示完成提示，如图 1-64 所示。

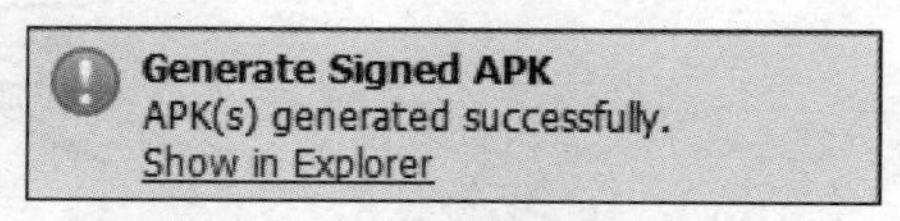

图 1-64　APK 完成提示

（7）在提示中单击 Show in Explorer 超链接，可打开文件资源管理器，查看 APK 文件，如图 1-65 所示。

图 1-65　在文件资源管理器中查看 APK 文件

这里的 app-release.apk 就是 Android Studio 生成的正式签名的 APK 文件。

1.6　小结

本章主要介绍了 Android 相关基础知识、如何搭建 Android 开发环境、如何创建 Android 项目以及如何在 Android 程序中使用日志工具。在将 Android Studio 作为 Android 应用开发环境时，应注意首先需要安装 JDK，然后安装 Android Studio。Google 在 Android Studio 安装程序中捆绑了 Android SDK。在使用 Android Studio 开发 Android 应用的过程中，可能会根据需要从网络下载不同版本的 Android SDK。Android Studio 在使用 Gradle 构建项目时，有时也会需要从网络下载组件，所以在开发过程中，保持网络连接很有必要。

1.7　习题

1. Android 平台具有哪些特点？
2. Android 平台体系架构可分为哪些层？
3. 在 Android Studio 中，可选择哪些设备来运行 Android 应用程序？
4. Android 中，注册活动、内容提供器等组件的文件名是什么？
5. 可用哪些对象来输出日志信息？

PART02

第2章

Android核心组件——活动

重点知识：

- 活动的基本操作
- 在活动中使用Intent
- 活动之间的数据传递
- 活动的生命周期
- 活动的启动模式

■ 第 1 章介绍了如何配置 Android 开发环境，并使用 Android Studio 创建了第一个 Android 程序。同时，也了解到 Android 程序启动时，总是会启动一个活动（Activity），然后将界面呈现给用户。可以说，Android 程序的设计总是从活动开始的。活动也是开发人员需要掌握的第一个 Android 核心组件。本章将详细介绍如何在 Android 程序中使用活动。

2.1 活动是什么

在使用 Android 手机时，无论是拨打电话、发送短信、浏览照片，还是运行其他应用，用户都会看到不同的界面。这些界面以及在界面中完成的各种操作，都通过活动完成。

活动是 Android 的一个核心应用组件，它主要用于实现应用功能逻辑，并通过界面显示数据或接收用户输入。一个应用程序可以包含零个或多个活动。如果应用程序没有活动，用户将无法看到程序界面，这种应用程序通常在后台运行，不涉及用户交互。

从用户的角度看，活动具有如下特点。

- 可通过返回键退出活动。
- 可通过 Home 键返回桌面。
- 可在活动中启动另一个界面，此时按返回键可返回前一个活动。

从开发人员的角度看，活动完成应用程序功能逻辑，它通过布局与用户交互，可以在活动中向另一个活动传递数据，也可接收另一个活动返回的数据。

一个应用通常包含多个活动，活动之间相对独立。包含多个活动的应用，需要为其指定一个“主”活动，即启动应用时首先打开的活动。每个活动均可启动另一个活动，包括当前活动本身。Android 甚至允许启动其他应用中的活动。

2.2 活动的基本操作

在 Android Studio 中添加活动时，如果使用了活动模板，模板可自动为活动添加视图。同时，Android Studio 还自动在清单文件 AndroidManifest.xml 中声明活动。本节主要介绍如何为活动绑定自定义视图、启动另一个活动、查看活动名称、退出活动以及销毁活动。

2.2.1 为活动绑定自定义视图

活动绑定自定义视图

通常，在活动的 onCreate()方法中使用 setContentView()方法来为活动绑定一个视图。

下面通过实例为活动绑定一个自定义视图，具体操作步骤如下。

（1）在 Android Studio 中选择“File\New\New Project”命令，打开新建项目对话框。将应用名称设置为 AndroidActivity，添加一个空活动，其他选项都使用默认值。

在选择“File\New\New Project”命令创建新项目时，Android Studio 会打开一个新对话框，原来打开的项目仍保持打开状态。选择“File\Close Project”命令可关闭当前项目，关闭项目后，Android Studio 会显示欢迎对话框，在欢迎对话框中选择“Start a new Android Studio project”选项创建新项目，这样就只有一个 Android Studio 窗口。

（2）在项目窗口（Android 模式）中，右击 res\layout 文件夹，在弹出的快捷菜单中选择“New\Layout resource file”命令，打开新资源文件对话框，如图 2-1 所示。

（3）将 File name 设置为 my_layout，其他选项使用默认值，单击 OK 按钮完成创建布局文件。Android Studio 会自动打开新建布局文件的布局编辑器的设计视图，如图 2-2 所示。在布局文件的设计视图中，不仅可预览当前布局，还可从 Palette 面板的 Widgets 列表中拖动组件到设计视图。布局编辑器窗口左下角有 Design 和 Text 两个选项卡，Design 选项卡显示布局可视化设计视图，Text 选项卡显示布局的 XML 文件编辑视图。

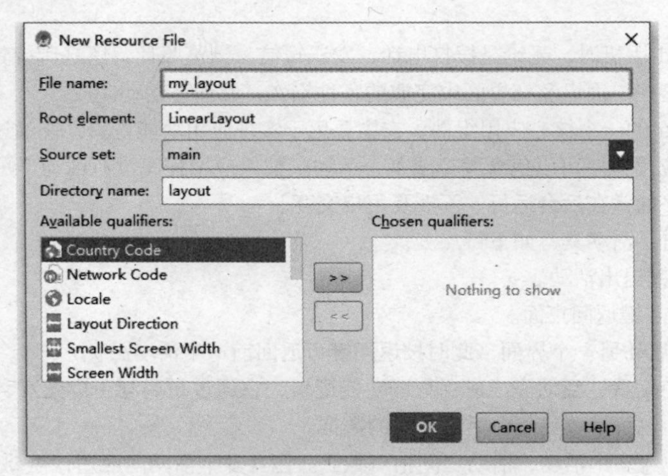

图 2-1　新建资源文件对话框

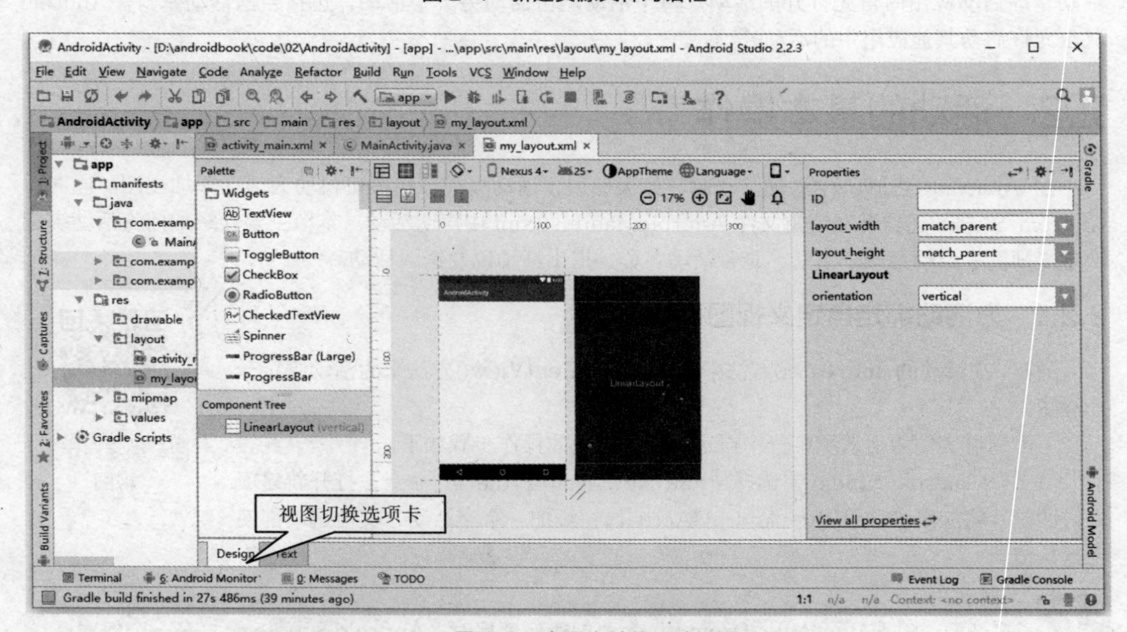

图 2-2　布局文件的设计视图

现在选择 Text 选项卡，可看到如下的布局文件代码。

```
<?xml version="1.0" encoding="utf-8"?>
<LinearLayout xmlns:android="http://schemas.android.com/apk/res/android"
    android:orientation="vertical" android:layout_width="match_parent"
    android:layout_height="match_parent">

</LinearLayout>
```

（4）修改布局文件 XML 代码，为布局添加一个按钮，代码如下。

```
<?xml version="1.0" encoding="utf-8"?>
```

```
<LinearLayout xmlns:android="http://schemas.android.com/apk/res/android"
    android:orientation="vertical" android:layout_width="match_parent"
    android:layout_height="match_parent">

    <Button
        android:text="Button"
        android:layout_width="match_parent"
        android:layout_height="wrap_content"
        android:id="@+id/button1" />
</LinearLayout>
```

在布局文件设计视图中，从 Palette 面板的 Widgets 列表中可拖动一个 Button 组件到设计视图，为布局添加一个 Button 组件。在 Text 视图中编辑代码时，可只输入关键字包含的几个字符，Android Studio 可自动显示匹配的关键字列表，提高了代码输入效率。在编写 Java 代码时，如果输入的关键字有错或需要添加类的导入（Import）时，单击红色显示的有误代码，再按【Alt+Enter】组合键即可获得提示，从提示中可选择完成导入的方法或其他更正错误的方法。

（5）打开活动源代码文件 MainActivity.java，修改 onCreate()方法，为活动绑定 my_layout 布局。onCreate()方法代码如下。

```
protected void onCreate(Bundle savedInstanceState) {
        super.onCreate(savedInstanceState);
        setContentView(R.layout.my_layout);                    //为活动绑定自定义视图
}
```

（6）按【Shift+F10】组合键运行项目，运行结果如图 2-3 所示。

图 2-3　绑定自定义视图后的运行结果

启动另一个 Activity

2.2.2 启动另一个活动

Android 应用在不同界面间切换时，其实质是在一个活动中启动了另一个活动，启动活动使用的是 startActivity()方法。

下面通过实例说明如何在一个活动中启动另一个活动，具体操作步骤如下。

（1）在 Android Studio 中创建一个项目，将应用名称设置为 StartAnotherActivity，添加一个 Empty Activity，其他选项为默认值。

（2）修改项目文件夹 app\res\layout 中的 activity_main.xml 文件，主要代码如下。

```
<?xml version="1.0" encoding="utf-8"?>
<RelativeLayout xmlns:android="http://schemas.android.com/apk/res/android"
...>


    <TextView
        android:layout_width="wrap_content"
        android:layout_height="wrap_content"
        android:text="这是主Activity"
        android:id="@+id/textView" />

    <Button
        android:text="启动另一个Activity"
        android:layout_width="wrap_content"
        android:layout_height="wrap_content"
        android:layout_below="@+id/textView"
        android:id="@+id/btnStartAnother" />
</RelativeLayout>
```

代码中修改了文本视图显示的字符串，并为布局添加了一个按钮，该按钮将用于打开另一个活动。

（3）在项目窗格中右击 app\java 文件夹中的包名，在打开的菜单中选择“New\Activity\Empty Activity”命令，所有参数使用默认值。

（4）修改新添加的布局文件 activity_main2.xml，添加一个文本视图，代码如下。

```
<?xml version="1.0" encoding="utf-8"?>
<RelativeLayout xmlns:android="http://schemas.android.com/apk/res/android"
    ...>
    <TextView
        android:layout_width="wrap_content"
        android:layout_height="wrap_content"
        android:text="这是另一个Activity"
        android:id="@+id/textView" />
</RelativeLayout>
```

（5）修改 MainActivity.java，为 btnStartAnother 按钮添加单击事件监听器，实现在单击按钮时启动另一个活动，代码如下。

```
package com.example.xbg.startanotheractivity;
import android.content.Intent;
import android.support.v7.app.AppCompatActivity;
import android.os.Bundle;
```

```
import android.view.View;
import android.widget.Button;
public class MainActivity extends AppCompatActivity {
    @Override
    protected void onCreate(Bundle savedInstanceState) {
        super.onCreate(savedInstanceState);
        setContentView(R.layout.activity_main);
        Button btn=(Button)findViewById(R.id.btnStartAnother);//引用布局中的按钮
        btn.setOnClickListener(new View.OnClickListener() {//添加单击事件监听器
            @Override
            public void onClick(View v) {
                startActivity(new Intent(MainActivity.this,Main2Activity.class));//启动另一个活动
            }
        });
    }
}
```

代码中，语句“startActivity(new Intent(MainActivity.this, Main2Activity.class));”用于启动另一个活动，其中 MainActivity 为当前活动 ID，Main2Activity 为另一个活动 ID。

（6）运行项目，查看运行效果。项目运行时，首先显示主活动界面，如图 2-4 所示，单击 启动另一个ACTIVITY 按钮，即可启动另一个活动，如图 2-5 所示。启动另一个活动后，按返回键可返回主活动界面。

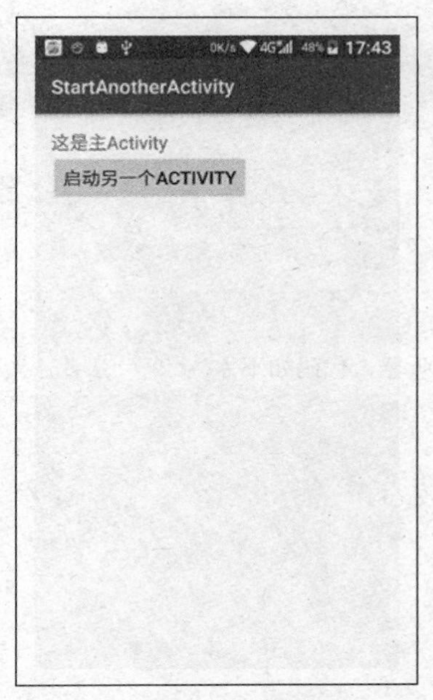

图 2-4　主活动界面

图 2-5　另一个活动界面

2.2.3　结束活动

在打开一个活动界面后，按返回键时，当前活动被结束。如果是按主页键，将返回设备主界面，此时活动在后台被挂起。

那么，如何主动结束活动呢？Activity 类提供了 finish()方法，用于结束当前活动。调用

finishActivity()方法可结束之前启动的其他活动。

下面通过实例说明如何结束一个活动，具体操作步骤如下。

（1）创建一个项目，将应用名称设置为 FinishActivity，为项目添加一个空活动，其他选项使用默认值。

（2）修改 activity_main.xml，为布局添加一个按钮，主要代码如下。

```
<?xml version="1.0" encoding="utf-8"?>
<RelativeLayout xmlns:android="http://schemas.android.com/apk/res/android"
    xmlns:tools="http://schemas.android.com/tools"
    android:id="@+id/activity_main"
    android:layout_width="match_parent"
    android:layout_height="match_parent"
    android:paddingBottom="@dimen/activity_vertical_margin"
    android:paddingLeft="@dimen/activity_horizontal_margin"
    android:paddingRight="@dimen/activity_horizontal_margin"
    android:paddingTop="@dimen/activity_vertical_margin"
    tools:context="com.example.xbg.finishactivity.MainActivity">

    <TextView
        android:layout_width="wrap_content"
        android:layout_height="wrap_content"
        android:text="Hello World!"
        android:id="@+id/textView" />

    <Button
        android:text="销毁活动"
        android:layout_width="wrap_content"
        android:layout_height="wrap_content"
        android:layout_below="@+id/textView"
        android:id="@+id/button" />
</RelativeLayout>
```

（3）修改 MainActivity.java，添加按钮单击事件监听器，代码如下。

```
package com.example.xbg.finishactivity;
import android.support.v7.app.AppCompatActivity;
import android.os.Bundle;
import android.view.View;
public class MainActivity extends AppCompatActivity {
    @Override
    protected void onCreate(Bundle savedInstanceState) {
        super.onCreate(savedInstanceState);
        setContentView(R.layout.activity_main);
        findViewById(R.id.button).setOnClickListener(new View.OnClickListener() {
            @Override
            public void onClick(View v) {
                finish();//结束活动
            }
```

```
        });
    }
}
```

（4）运行项目，查看运行效果。项目运行效果如图 2-6 所示，单击 结束活动 按钮，可关闭活动，与按返回键的效果相同。

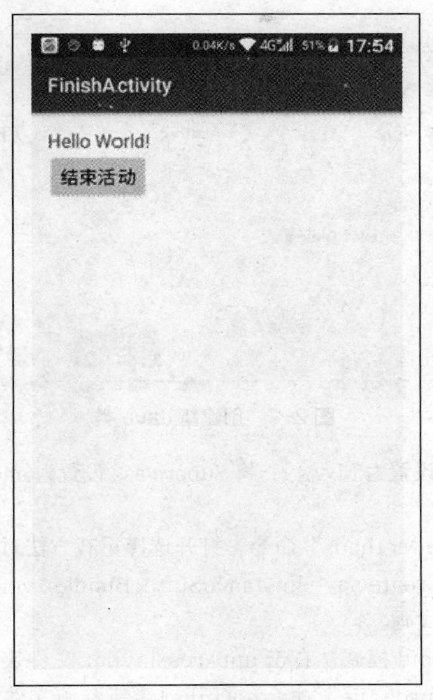

图 2-6　结束活动

一般情况下，不需要主动调用 finish()方法来结束一个活动，因为活动结束后将无法返回该活动。只有在确实不想让用户返回时，才需要调用 finish()方法结束活动。

2.3　在活动中使用 Intent

Intent 是 Android 应用中的一种消息传递机制，通过 Intent 对象实现应用组件之间的通信。通常，Intent 用于启动活动、启动服务以及发送广播。Intent 可分为显式 Intent 和隐式 Intent。

2.3.1　显式 Intent

显式 Intent 指在创建 Intent 对象时，指定了要启动的特定组件。

2.2.2 节中已经使用了显式 Intent 来启动另一个活动。下面通过实例创建一个自定义的类来实现一个活动，并使用显式 Intent 来启动该活动，具体操作步骤如下。

显式 Intent

（1）在 Android Studio 中创建一个新项目，将应用名称设置为 LearnIntent，并为项目添加一个空活动。

（2）在项目窗格中（Android 模式）右击 app\java 文件夹中的包名称，在弹出的快捷菜单中选择“New\Java Class”命令，打开创建新 Java 类对话框，如图 2-7 所示。

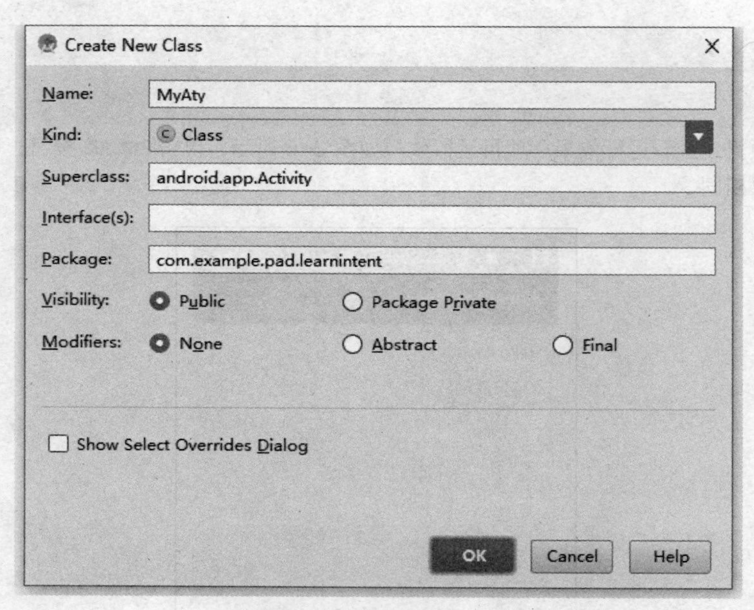

图 2-7　创建新 Java 类

（3）将新 Java 类的 Name 设置为 MyAty，将 Superclass 设置为 android.app.Activity，单击 OK 按钮完成创建新类。

（4）选择“Code\Override Methods”命令，打开选择重载方法对话框，如图 2-8 所示。

（5）在列表框中双击 onCreate(savedInstanceState:Bundle):void，选择重载该方法，Android Studio 会自动添加方法的重载代码。

（6）在项目窗格中（Android 模式）右击 app\res\layout 文件夹中的包名称，在弹出的快捷菜单中选择“New\Layout resource file”命令，打开创建新资源文件对话框，将新的布局文件命名为 myaty。

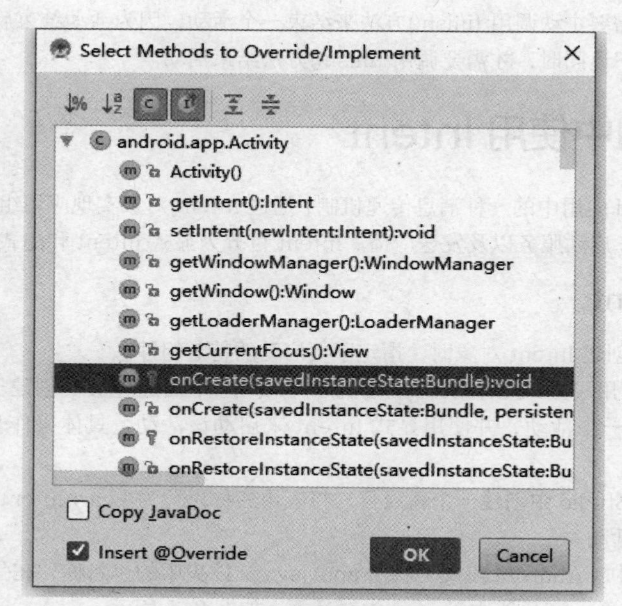

图 2-8　选择重载的方法

（7）在设计视图中为布局添加一个 TextView，并将其 text 属性设置为“这是使用显式 Intent 打开的活动”。

（8）返回 MyAty.java 代码编辑窗口，修改 onCreate()，为活动绑定刚创建的布局，代码如下。

```
package com.example.pad.learnintent;
import android.app.Activity;
import android.os.Bundle;
public class MyAty extends Activity {
    @Override
    protected void onCreate(Bundle savedInstanceState) {
        super.onCreate(savedInstanceState);
        setContentView(R.layout.myaty);
    }
}
```

（9）打开 activity_main.xml 设计视图，为布局添加一个 Button 组件，将组件 ID 属性设置为 btnStartMyAty，将 text 属性设置为“启动 MYATY”。

（10）打开 MainActivity.java 代码编辑窗口，修改 onCreate()方法，添加按钮单击事件监听器，在单击按钮时启动活动 MyAty，代码如下。

```
package com.example.pad.learnintent;
import android.content.Intent;
import android.support.v7.app.AppCompatActivity;
import android.os.Bundle;
import android.view.View;
public class MainActivity extends AppCompatActivity {
    @Override
    protected void onCreate(Bundle savedInstanceState) {
        super.onCreate(savedInstanceState);
        setContentView(R.layout.activity_main);
        findViewById(R.id.btnStartMyAty).setOnClickListener(new View.OnClickListener() {
            @Override
            public void onClick(View v) {
                Intent startMyAty=new Intent(MainActivity.this,MyAty.class);
                startActivity(startMyAty);
            }
        });
    }
}
```

代码中，语句 new Intent(MainActivity.this,MyAty.class)创建了一个 Intent 对象，传递的第一个参数 MainActivity.this 指定启动活动的上下文为当前活动，第二个参数 MyAty.class 指定启动的目标组件为 MyAty。这种方式创建的 Intent 对象就是典型的显式 Intent。

（11）按【Shift+F10】组合键运行项目，项目主界面如图 2-9 所示。

（12）单击 启动MYATY 按钮，尝试启动活动 MyAty，应用运行出错，如图 2-10 所示。这是因为自定义的活动 MyAty 还没有在 AndroidManifest.xml 中声明。

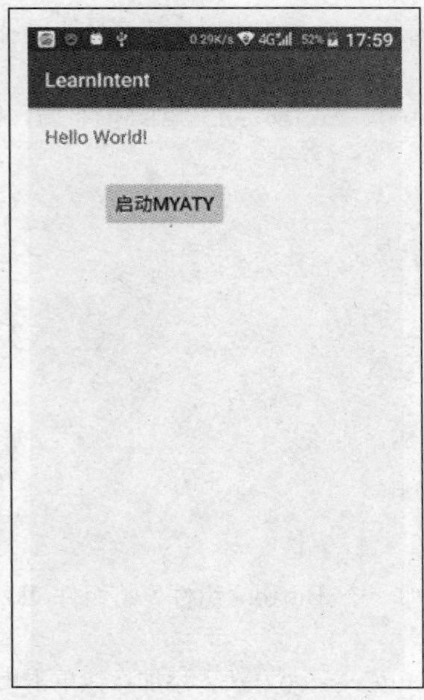

图 2-9　项目主界面

图 2-10　运行出错

在 Android Studio 的 Run 窗口（按【Alt+F4】组合键可打开 Run 窗口）中，可看到如下提示信息。

```
D/AndroidRuntime: Shutting down VM
E/AndroidRuntime: FATAL EXCEPTION: main
                  Process: com.example.pad.learnintent, PID: 23904
                  android.content.ActivityNotFoundException: Unable to find explicit activity class {com.example.pad.learnintent/com.example.pad.learnintent.MyAty}; have you declared this activity in your AndroidManifest.xml?
                      at android.app.Instrumentation.checkStartActivityResult(Instrumentation.java:1805)
...
I/Process: Sending signal. PID: 23904 SIG: 9
Application terminated.
```

信息中说明了应用程序找不到活动 com.example.pad.learnintent.MyAty，并提示应该在 AndroidManifest.xml 中声明该活动。

（13）修改 AndroidManifest.xml，添加活动 MyAty 的声明信息，代码如下。

```
<?xml version="1.0" encoding="utf-8"?>
<manifest xmlns:android="http://schemas.android.com/apk/res/android"
    package="com.example.pad.learnintent">
    <application
        android:allowBackup="true"
        android:icon="@mipmap/ic_launcher"
        android:label="@string/app_name"
        android:supportsRtl="true"
        android:theme="@style/AppTheme">
        <activity android:name=".MainActivity">
            <intent-filter>
                <action android:name="android.intent.action.MAIN" />
```

```
                <category android:name="android.intent.category.LAUNCHER" />
            </intent-filter>
        </activity>
        <activity android:name=".MyAty"/>
    </application>
</manifest>
```

代码中，<activity android:name=".MyAty"/>为活动声明语句，只需指明活动的名称即可。注意，MyAty 名称前面的点号表示当前包名，如果要使用完整的名称，则应该是用语句 com.example.pad.learnintent.MyAty。

（14）按【Shift+F10】组合键，重新运行项目。在主界面中单击启动MYATY按钮，可正确启动活动 MyAty，活动 MyAty 运行界面如图 2-11 所示。

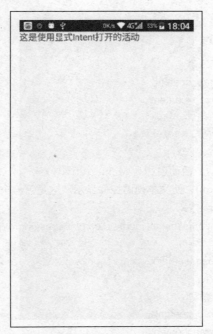

图 2-11　活动 MyAty 运行界面

本节实例除了说明了如何使用显式 Intent 外，还说明了一个项目中的所有活动都应该在项目清单文件 AndroidManifest.xml 中进行声明。在 2.2.2 小节中为项目添加 Activity 时，Android Studio 自动添加了活动的声明信息。本节通过自定义 Java 类来实现活动，所以需要手动添加活动声明信息。

2.3.2　隐式 Intent

隐式 Intent

显式 Intent 指明了要启动的组件，隐式 Intent 则相反，它不指明要启动的组件，而是指明要执行的操作，让系统去选择可完成该操作的组件。

1. 启动同一个应用中的活动

下面通过一个实例说明如何使用隐式 Intent 启动同一个应用中的活动，具体操作步骤如下。

（1）在 Android Studio 中创建一个新项目，将应用名称设置为 LearnHideIntent1，并为项目添加一个空活动。

（2）修改 activity_main.xml，在主活动布局中添加一个按钮控件，代码如下。

```
<?xml version="1.0" encoding="utf-8"?>
<RelativeLayout xmlns:android="http://schemas.android.com/apk/res/android"
...
    tools:context="com.example.pad.learnhideintent1.MainActivity">

    <TextView    android:id="@+id/textView"
        android:layout_width="wrap_content"
        android:layout_height="wrap_content"
        android:text="主活动" />

    <Button
        android:text="启动另一个活动"
        android:layout_width="wrap_content"
        android:layout_height="wrap_content"
        android:layout_below="@+id/textView"
        android:id="@+id/button" />
</RelativeLayout>
```

（3）为项目添加一个空活动，活动名称为 Main2Activity。

（4）修改 activity_main2.xml，使文本视图控件显示"这是另一个活动"，代码如下。

```
<?xml version="1.0" encoding="utf-8"?>
<RelativeLayout xmlns:android="http://schemas.android.com/apk/res/android"
...
    tools:context="com.example.pad.learnhideintent1.Main2Activity">
    <TextView
        android:text="这是另一个活动"
        android:layout_width="wrap_content"
        android:layout_height="wrap_content"
        android:id="@+id/textView2" />
</RelativeLayout>
```

（5）修改 AndroidManifest.xml，添加活动 Main2Activity 的 Intent 过滤器，代码如下。

```
<?xml version="1.0" encoding="utf-8"?>
<manifest xmlns:android="http://schemas.android.com/apk/res/android"
    package="com.example.pad.learnhideintent1">
    <application
        android:allowBackup="true"
        android:icon="@mipmap/ic_launcher"
        android:label="@string/app_name"
        android:supportsRtl="true"
        android:theme="@style/AppTheme">
        <activity android:name=".MainActivity">
            <intent-filter>
```

```
                <action android:name="android.intent.action.MAIN" />
                <category android:name="android.intent.category.LAUNCHER" />
            </intent-filter>
        </activity>
        <activity android:name=".Main2Activity">
            <intent-filter>
                <category android:name="android.intent.category.DEFAULT"/>
                <action android:name="toStartAnotherActivity"/>
            </intent-filter>
        </activity>
    </application>
</manifest>
```

代码中，活动 Main2Activity 的 Intent 过滤器中定义了一个操作 toStartAnotherActivity，该名称将作为参数用于创建隐式 Intent，以启动活动 Main2Activity。

（6）修改 MainActivity.java，添加按钮单击事件监听器，在单击按钮时，使用隐式 Intent 启动活动 Main2Activity，代码如下。

```
package com.example.pad.learnhideintent1;
import android.content.Intent;
import android.support.v7.app.AppCompatActivity;
import android.os.Bundle;
import android.view.View;
public class MainActivity extends AppCompatActivity {
    @Override
    protected void onCreate(Bundle savedInstanceState) {
        super.onCreate(savedInstanceState);
        setContentView(R.layout.activity_main);
        findViewById(R.id.button).setOnClickListener(new View.OnClickListener() {
            @Override
            public void onClick(View v) {
                startActivity(new Intent("toStartAnotherActivity"));
            }
        });
    }
}
```

代码中，语句 new Intent("toStartAnotherActivity")创建了一个隐式 Intent，参数为一个操作名称。项目运行时，系统用该名称到清单文件中搜索匹配的活动。

（7）运行项目，测试运行效果。图 2-12 所示为项目主活动界面，单击启动另一个活动按钮可打开另一个活动界面，如图 2-13 所示。

通常在 AndroidManifest.xml 中声明活动时，在其 Intent 过滤器中可使用任意字符串作为活动的操作（action）的名称。作为一种编码习惯，推荐使用活动的完全限定类名称作为操作名称，例如 com.example.pad.learnhideintent1.Main2Activity，这样代码更易于阅读。

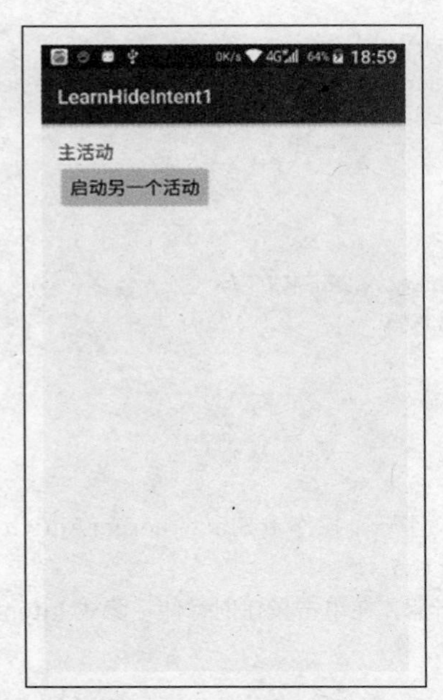

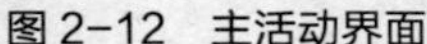
图 2-12　主活动界面

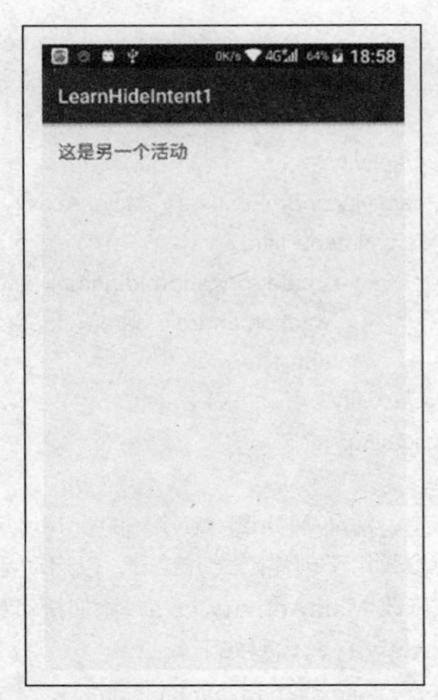

图 2-13　另一个活动界面

2．启动另一个应用中的活动

在使用显式 Intent 启动活动时，系统可直接启动 Intent 对象中指定的组件。而在使用隐式 Intent 时，系统则会搜索设备上所有应用的清单文件中的 Intent 过滤器，找到匹配的操作时，系统启动对应的组件。所以，使用隐式 Intent 可启动另一个应用中的活动。

下面通过实例说明如何启动另一个项目中的活动，具体操作步骤如下。

（1）在 Android Studio 中创建一个新项目，将应用名称设置为 LearnHideIntent2，并为项目添加一个空活动。

（2）修改 activity_main.xml，在主活动布局中添加一个按钮控件，主要代码如下。

```
<?xml version="1.0" encoding="utf-8"?>
<RelativeLayout xmlns:android="http://schemas.android.com/apk/res/android"
    ...>
    <TextView
        android:layout_width="wrap_content"
        android:layout_height="wrap_content"
        android:text="Hello World!"
        android:id="@+id/textView" />
    <Button
        android:text="启动另一个应用中的活动"
        android:layout_width="wrap_content"
        android:layout_height="wrap_content"
        android:layout_below="@+id/textView"
        android:id="@+id/button" />
</RelativeLayout>
```

（3）修改 MainActivity.java，添加按钮单击事件监听器，在单击按钮时，使用隐式 Intent 启动活动 Main2Activity，代码如下。

```
package com.example.pad.learnhideintent2;
import android.content.Intent;
```

```
import android.support.v7.app.AppCompatActivity;
import android.os.Bundle;
import android.view.View;
public class MainActivity extends AppCompatActivity {
    @Override
    protected void onCreate(Bundle savedInstanceState) {
        super.onCreate(savedInstanceState);
        setContentView(R.layout.activity_main);
        findViewById(R.id.button).setOnClickListener(new View.OnClickListener() {
            @Override
            public void onClick(View v) {
                startActivity(new Intent("toStartAnotherActivity"));
            }
        });
    }
}
```

（4）运行项目，测试运行效果。图 2-14 所示为项目主活动界面，单击启动另一个应用中的活动按钮可打开 LearnHideIntent1 中的活动 Main2Activity，其运行效果与图 2-13 相同。

默认情况下，一个应用中的所有活动均可被其他应用调用。如果要限制活动只能在当前活动中使用，则可在清单文件中声明活动时将 android:exported 属性设置为 false（其默认值为 true），表示不允许导出。例如如下代码。

```
<activity android:name=".Main2Activity" android:exported="false">
    <intent-filter>
        <category android:name="android.intent.category.DEFAULT" />
        <action android:name="toStartAnotherActivity" />
    </intent-filter>
</activity>
```

如果在 LearnHideIntent1 的清单文件中禁止导出 Main2Activity，则在运行 LearnHideIntent2 时将无法启动活动 Main2Activity，错误提示如图 2-15 所示。

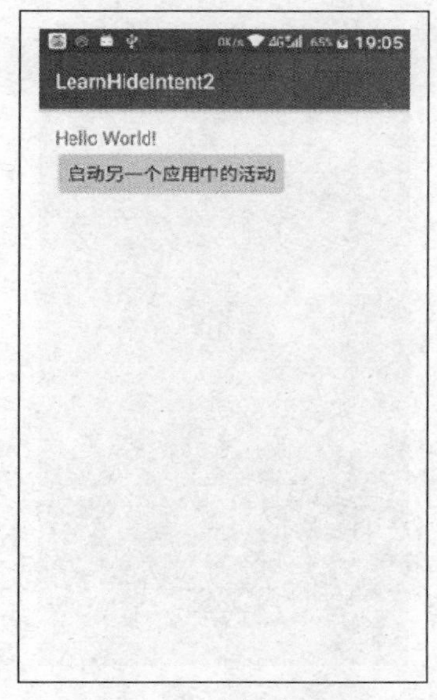

图 2-14　主活动界面

图 2-15　不能启动活动 Main2Activity 时的错误提示

3. 使用预定义操作

在创建隐式 Intent 时，将清单文件中声明的活动的操作名称作为参数。Android 提供了一系列预定义操作，使用这些操作可启动 Android 内置应用，例如打开浏览器、编辑联系人、拨打电话等。

使用隐式 Intent 执行预定义操作时，需要为 Intent 对象指定动作和数据。常用预定义动作如下。

- ACTION_VIEW：打开一个视图，显示指定的数据。例如，数据为 URL 时会打开浏览器，数据为 content://contacts/people/1 时显示第一个联系人信息，数据为 content://contacts/people/时显示联系人列表，数据为 tel:10000 时显示拨打电话界面。
- ACTION_DIAL：显示拨打电话界面，数据格式为“tel:电话号码”，与 ACTION_VIEW 类似。
- ACTION_EDIT：显示数据编辑界面，例如，数据为 content://contacts/people/1 时，显示第一个联系人信息的编辑界面。

相同的操作在不同设备中运行时，运行界面可能会有所区别。这是因为不同设备中用来完成这一操作的组件可能会有所不同。

要创建隐式 Intent 执行预定义操作，需要先创建执行指定操作的 Intent 对象，然后调用对象的 setData()方法设置数据，例如如下代码。

```
Intent intent=new Intent(Intent.ACTION_VIEW);
intent.setData(Uri.parse("http://www.jikexueyuan.com"));
```

setData()方法要求参数为 URI 对象，所以调用了 parse()方法来构造 URI 对象。

下面通过实例说明如何使用隐式 Intent 执行预定义操作来打开浏览器、编辑联系人和拨打电话等，具体操作步骤如下。

（1）在 Android Studio 中创建一个新项目，将应用名称设置为 LearnHideIntent3，并为项目添加一个空活动。

（2）修改 activity_main.xml，在主活动布局中添加 3 个按钮控件，代码如下。

```
<?xml version="1.0" encoding="utf-8"?>
<LinearLayout xmlns:android="http://schemas.android.com/apk/res/android"
    ...>
    <TextView
        android:layout_width="wrap_content"
        android:layout_height="wrap_content"
        android:text="使用Android预定义操作" />
    <Button
        android:text="打开浏览器"
        android:layout_width="wrap_content"
        android:layout_height="wrap_content"
        android:id="@+id/button1"      />
    <Button
        android:text="编辑联系人"
        android:layout_width="wrap_content"
        android:layout_height="wrap_content"
        android:id="@+id/button2" />
    <Button
        android:text="拨打电话"
        android:layout_width="wrap_content"
```

```
        android:layout_height="wrap_content"
        android:id="@+id/button3"/>
</LinearLayout>
```

（3）修改 MainActivity.java，分别为按钮控件添加 3 个按钮的单击事件监听器，代码如下。

```
package com.example.pad.learnhideintent3;
import android.content.Intent;
...
public class MainActivity extends AppCompatActivity {
    @Override
    protected void onCreate(Bundle savedInstanceState) {
        super.onCreate(savedInstanceState);
        setContentView(R.layout.activity_main);
        findViewById(R.id.button1).setOnClickListener(new View.OnClickListener() {
            @Override
            public void onClick(View v) {
                    //使用隐式Intent打开浏览器
                    Intent intent = new Intent(Intent.ACTION_VIEW);
                    intent.setData(Uri.parse("http://www.jikexueyuan.com"));
                    startActivity(intent);
            }
        });
        findViewById(R.id.button2).setOnClickListener(new View.OnClickListener() {
            public void onClick(View v) {
                //使用隐式Intent打开联系人信息编辑界面
                Intent intent=new Intent(Intent.ACTION_EDIT);
                intent.setData(Uri.parse("content://contacts/people/1"));
                startActivity(intent);
            }
        });
        findViewById(R.id.button3).setOnClickListener(new View.OnClickListener() {
            @Override
            public void onClick(View v) {
                //使用隐式Intent打开拨打电话界面
                Intent intent=new Intent(Intent.ACTION_DIAL);
                intent.setData(Uri.parse("tel:87721234"));
                startActivity(intent);
            }
        });
    }
}
```

（4）运行项目，测试运行效果。

项目运行时，主活动界面如图 2-16 所示，单击“打开浏览器”按钮，可打开浏览器，显示指定的网页。因为测试手机中有多种浏览器可用，所以会显示图 2-17 所示的提示对话框，这说明系统在执行隐式 Intent 时将会搜索设备中所有可能执行该操作的应用，将其清单提供给用户选择。在提示对话框中，若选择“仅此一次”，表示只在本次操作时用选择的应用执行操作；若选择“始终”，则表示将当前选择设置为默认，以后再执行相同操作时不会再弹出提示。如果想重新显示提示，则需要在系统设置中打开执行操作时选择应用的应用信息设置，清除其默认设置。

图 2-18 所示为在浏览器中打开的网页，此时按返回键可返回主活动界面。在主活动界面单击“编辑联系人”按钮，可打开设备中第一个联系人信息的编辑界面，如图 2-19 所示。

在主活动界面单击“拨打电话”按钮时，可打开拨打电话界面，并显示指定的电话号码，如图 2-20 所示。

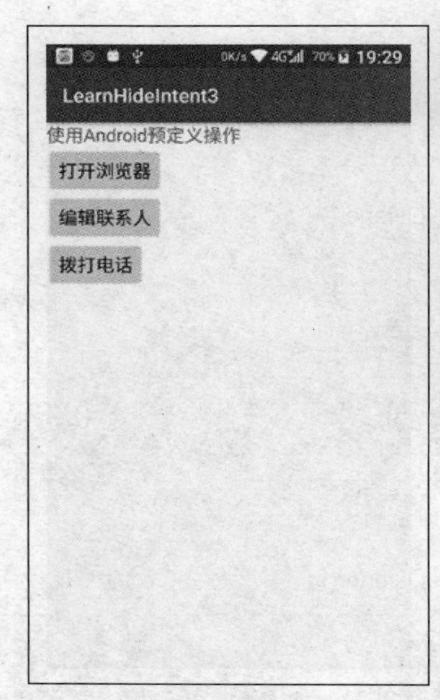

图 2-16　主活动界面

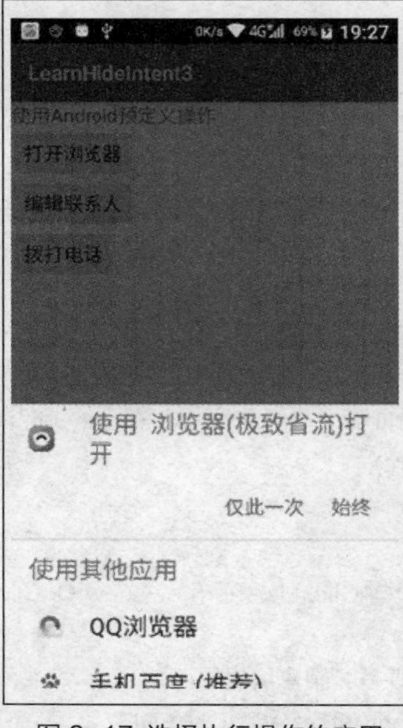

图 2-17　选择执行操作的应用

图 2-18　打开浏览器

图 2-19　第一个联系人信息编辑界面

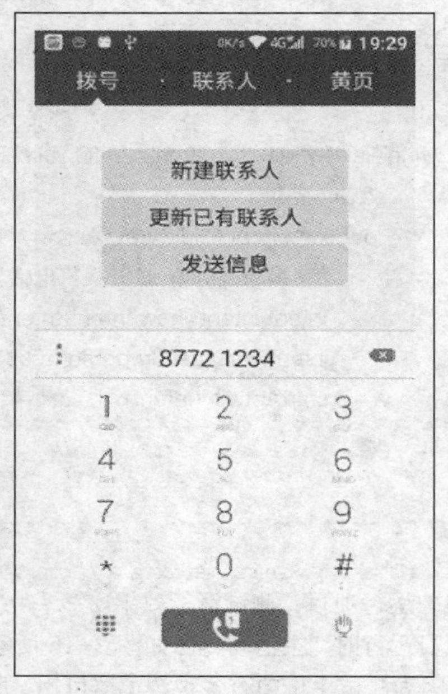

图 2-20　拨打电话界面

2.3.3 Intent 过滤器

Intent 过滤器

Intent 过滤器主要用于声明应用组件可接收的 Intent 操作、数据和其他设置。

前文已多次使用了 Intent 过滤器，每个应用主活动的声明与如下代码类似。

```
<activity android:name=".MainActivity">
    <intent-filter>
        <action android:name="android.intent.action.MAIN" />
        <category android:name="android.intent.category.LAUNCHER" />
    </intent-filter>
</activity>
```

上述代码中，<intent-filter>元素为活动 MainActivity 声明了一个过滤器；<action>元素声明活动可接受的操作为 android.intent.action.MAIN，表示当前活动作为应用的主入口点；<category>元素声明类别为 android.intent.category.LAUNCHER，即表示当前活动作为最顶层的启动器。操作和类别结合，表示启动应用时首先启动当前活动。

清单文件 AndroidManifest.xml 声明 Intent 过滤器时，可使用如下 3 个元素。

- <action>元素：在其 name 属性中声明组件可接受的 Intent 操作，操作名称可以是自定义的文本字符串或者 android.intent.action 类的常量。
- <category>元素：在其 name 属性中声明组件可接受的 Intent 类别，类别名称通常为 android.intent.category 类中的常量。如果要让活动响应隐式 Intent，则必须将过滤器的类别设置为 android.intent.category.DEFAULT。如果没有在 Intent 过滤器中声明 DEFAULT 类别，则隐式 Intent 不会解析该组件。
- <data>元素：声明数据 URI 的 scheme、host、port、path 等，或者是数据的 MIME 类型。

在代码中创建 Intent 对象时，可调用下列方法为 Intent 对象添加操作、类别、数据或其他属性。

- setAction()：设置 Intent 对象操作。也可在 Intent 对象构造函数中指定操作。
- addCategory()：为 Intent 对象添加类别。
- setData()：设置数据 URI。
- setType()：设置 MIME 类型。
- setDataAndType()：setData()和 setType()会相互抵消彼此的设置。若要同时设置 URI 和 MIME 类型，则需调用 setDataAndType()。

下面通过实例进一步说明 Intent 过滤器的使用，具体操作步骤如下。

（1）在 Android Studio 中创建一个新项目，将应用名称设置为 LearnIntentFilter，并为项目添加一个空活动。

（2）修改 activity_main.xml，在主活动布局中添加一个按钮控件，主要代码如下。

```
<?xml version="1.0" encoding="utf-8"?>
<RelativeLayout xmlns:android="http://schemas.android.com/apk/res/android"
    ...>
    <TextView
        android:layout_width="wrap_content"
        android:layout_height="wrap_content"
        android:text="Hello World!"
```

```
        android:id="@+id/textView" />
    <Button
        android:text="启动活动"
        android:layout_width="wrap_content"
        android:layout_height="wrap_content"
        android:id="@+id/button"
        android:layout_below="@id/textView" />
</RelativeLayout>
```

（3）为项目添加一个空活动，修改其布局文件，其主要代码如下。

```
<?xml version="1.0" encoding="utf-8"?>
<RelativeLayout xmlns:android="http://schemas.android.com/apk/res/android"
    ...>
    <TextView
        android:text="Main2Activity"
        android:layout_width="wrap_content"
        android:layout_height="wrap_content"
        android:id="@+id/textView2" />
</RelativeLayout>
```

文本视图控件可以显示当前活动的名称，以便和另一个活动进行区别。

（4）再为项目添加一个空活动，修改其布局文件，其主要代码如下。

```
<?xml version="1.0" encoding="utf-8"?>
<RelativeLayout xmlns:android="http://schemas.android.com/apk/res/android"
    ...>
    <TextView
        android:text="Main3Activity"
        android:layout_width="wrap_content"
        android:layout_height="wrap_content"
        android:id="@+id/textView3" />
</RelativeLayout>
```

（5）修改清单文件 AndroidManifest.xml，为前面添加的两个活动添加声明信息，代码如下。

```
<?xml version="1.0" encoding="utf-8"?>
<manifest xmlns:android="http://schemas.android.com/apk/res/android"
    package="com.example.pad.learnintentfilter">
    <application
        ...>
        <activity android:name=".MainActivity">
            <intent-filter>
                <action android:name="android.intent.action.MAIN" />
                <category android:name="android.intent.category.LAUNCHER" />
            </intent-filter>
        </activity>
        <activity android:name=".Main2Activity" android:label="Main2Activity">
```

```
            <intent-filter>
                <category android:name="android.intent.category.DEFAULT" />
                <action android:name="toDoSomething" />
            </intent-filter>
        </activity>
        <activity android:name=".Main3Activity" android:label="Main3Activity">
            <intent-filter>
                <category android:name="android.intent.category.DEFAULT" />
                <action android:name="toDoSomething" />
            </intent-filter>
        </activity>
    </application>
</manifest>
```

代码中，活动 Main2Activity 和 Main3Activity 的操作名称都设置为 toDoSomething，在隐式 Intent 中使用 toDoSomething 作为操作名称时，系统会显示对话框让用户选择打开哪一个活动，android:label 属性值用于在提示对话框中用于区别两个活动。

（6）修改 MainActivity.java，添加按钮的单击事件监听器，在单击按钮时，用隐式 Intent 启动活动，主要代码如下。

```
package com.example.pad.learnintentfilter;
import android.content.Intent;
…
public class MainActivity extends AppCompatActivity {

    @Override
    protected void onCreate(Bundle savedInstanceState) {
        super.onCreate(savedInstanceState);
        setContentView(R.layout.activity_main);
        findViewById(R.id.button).setOnClickListener(new View.OnClickListener() {
            @Override
            public void onClick(View v) {
                Intent intent=new Intent();
                intent.setAction("toDoSomething");
                startActivity(intent);
            }
        });
    }
}
```

（7）运行项目，测试运行结果。

项目启动后显示的主活动界面如图 2-21 所示，单击“启动活动”按钮，会打开图 2-22 所示的对话框，提示选择要打开的活动。

如果选择 Main2Activity，则会打开图 2-23 所示的界面。如果选择 Main3Activity，则会打开图 2-24 所示的界面。

图 2-21　主活动界面

图 2-22　提示对话框

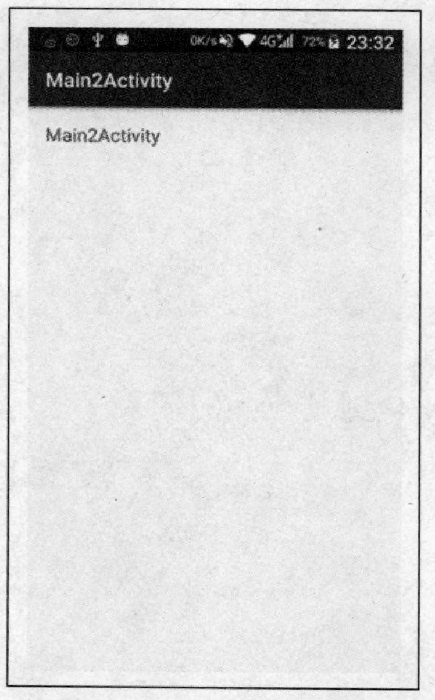

图 2-23　Main2Activity 界面

图 2-24　Main3Activity 界面

（8）停止项目运行，再修改清单文件 AndroidManifest.xml，为 Main2Activity 添加一个数据，代码如下。

```
<activity android:name=".Main2Activity"   android:label="Main2Activity">
    <intent-filter>
        <category android:name="android.intent.category.DEFAULT" />
        <action android:name="toDoSomething" />
        <data android:scheme="app"/>
```

```
    </intent-filter>
</activity>
```

（9）修改 MainActivity.java 中的按钮单击事件监听器，为隐式 Intent 添加数据，代码如下。

```
protected void onCreate(Bundle savedInstanceState) {
        super.onCreate(savedInstanceState);
        setContentView(R.layout.activity_main);
        findViewById(R.id.button).setOnClickListener(new View.OnClickListener() {
            @Override
            public void onClick(View v) {
                Intent intent=new Intent();
                intent.setAction("toDoSomething");
                intent.setData(Uri.parse("app://anything"));
                startActivity(intent);
            }
        });
}
```

（10）再次运行项目，在主活动界面中单击“启动活动”按钮时，将不会再显示提示对话框，而是直接启动了 Main2Activity。

该实例说明 Intent 过滤器的作用就是帮助系统解析隐式 Intent，去选择要启动的组件。只有当 Intent 过滤器用的动作和数据与隐式 Intent 完全匹配时，才会启动该组件。如果有多个匹配的 Intent 过滤器，则会在提示对话框中显示活动清单，供用户选择。

2.3.4 从网页中启动活动

通过浏览器链接启动本地 Activity

Android 允许在浏览器中启动活动，Intent 过滤器中包含 BROWSABLE 类别，即表示当前活动可从浏览器启动。

下面通过实例说明如何创建一个可供浏览器启动的活动，并通过浏览器链接来启动，具体操作步骤如下。

（1）在 Android Studio 中创建一个新项目，将应用名称设置为 LaunchLocalApp，并为项目添加一个空活动。

（2）再为项目添加一个空活动，将活动名称设置为 LocalAppAty。

（3）修改 activity_local_app_aty.xml，为刚添加的活动增加一个文本视图控件，主要代码如下。

```
<?xml version="1.0" encoding="utf-8"?>
<RelativeLayout xmlns:android="http://schemas.android.com/apk/res/android"
   …>
    <TextView
        android:text="这是用于被浏览器链接启动的一个本地Activity"
        android:layout_width="wrap_content"
        android:layout_height="wrap_content"
        android:id="@+id/textView1" />
    <TextView
        android:text=""
        android:layout_width="wrap_content"
        android:layout_height="wrap_content"
        android:layout_below="@+id/textView1"
```

```
        android:id="@+id/textView2" />
</RelativeLayout>
```

textView1 用于显示活动的说明信息，textView2 用于显示启动活动时的 URI 信息。

（4）修改 LocalAppAty.java，在活动启动时将 URI 信息显示到 textView2 中，代码如下。

```
package com.example.pad.launchlocalapp;
import android.net.Uri;
…
public class LocalAppAty extends AppCompatActivity {
    @Override
    protected void onCreate(Bundle savedInstanceState) {
        super.onCreate(savedInstanceState);
        setContentView(R.layout.activity_local_app_aty);
        Uri uri=getIntent().getData();                              //获得URI字符串
        TextView txt=(TextView)findViewById(R.id.textView2);
        String suri=Uri.decode(uri.toString());                     //解码URI字符串
        String data=uri.getQueryParameter("data");                  //获得URI中的查询参数
        txt.setText("\n下面是URI和查询参数：\n"+suri+"\n查询参数data="+data);
    }
}
```

代码中对 URI 字符串进行了解码，如果不解码，其中的汉字将不能正常显示。

（5）修改 activity_main.xml，在主活动布局中添加一个按钮控件，代码如下。

```
<?xml version="1.0" encoding="utf-8"?>
<RelativeLayout xmlns:android="http://schemas.android.com/apk/res/android"
    …>

    <TextView
        android:layout_width="wrap_content"
        android:layout_height="wrap_content"
        android:text="Hello World!"
        android:id="@+id/textView" />
    <Button
        android:text="启动本地Activity"
        android:layout_width="wrap_content"
        android:layout_height="wrap_content"
        android:layout_below="@id/textView"
        android:id="@+id/button" />
</RelativeLayout>
```

按钮 button 用于在项目运行时在主活动中启动 LocalAppAty 来进行测试。

（6）修改清单文件 AndroidManifest.xml，为活动 LocalAppAty 添加声明信息，代码如下。

```
<?xml version="1.0" encoding="utf-8"?>
<manifest xmlns:android="http://schemas.android.com/apk/res/android"
    package="com.example.pad.launchlocalapp">
    <application …>
            <intent-filter>
                <action android:name="android.intent.action.MAIN" />
```

```
                <category android:name="android.intent.category.LAUNCHER" />
            </intent-filter>
        </activity>
        <activity android:name=".LocalAppAty" android:label="本地活动LocalAppAty">
            <intent-filter>
                <category android:name="android.intent.category.DEFAULT"/>
                <category android:name="android.intent.category.BROWSABLE"/>
                <action android:name="android.intent.action.VIEW"/>
                <data android:scheme="app"/>
            </intent-filter>
        </activity>
    </application>
</manifest>
```

在浏览器中，将使用<data android:scheme="app"/>声明的 app 协议来作为超链接地址。Android 系统浏览器中使用 app 协议时，会搜索设备中所有应用的清单文件，找到匹配的组件后，Android 系统会将 URI 封装到隐式 Intent 中，传递给组件，并启动该组件。设备中的浏览器其实也是一个 Android 应用，所以在其中启动本地活动，其过程和在本地应用中启动其他应用中的活动的过程相同。

要使活动响应浏览器，除了在 Intent 过滤器中包含 BROWSABLE 类别，还需包含 DEFAULT 类别（因为是通过隐式 Intent 启动活动）和 VIEW 活动（因为活动用于显示数据）。

（7）修改 MainActivity.java，添加按钮单击事件监听器，在主活动中启动 LocalAppAty 活动来测试其运行效果，代码如下。

```
package com.example.pad.launchlocalapp;
import android.content.Intent;
...
public class MainActivity extends AppCompatActivity {
    @Override
    protected void onCreate(Bundle savedInstanceState) {
        super.onCreate(savedInstanceState);
        setContentView(R.layout.activity_main);
        findViewById(R.id.button).setOnClickListener(new View.OnClickListener() {
            @Override
            public void onClick(View v) {
                Intent intent=new Intent(MainActivity.this,LocalAppAty.class);
                intent.setData(Uri.parse("本地启动测试"));
                startActivity(intent);//启动本地Activity，测试其运行效果
            }
        });
    }
}
```

（8）运行项目，测试其运行效果。

项目运行时，主活动界面如图 2-25 所示。单击按钮启动 LocalAppAty 活动，界面如图 2-26 所示。可以看到，在主活动中启动时，LocalAppAty 接收到的 Intent URI 中没有查询参数 data，所以其值为 null。

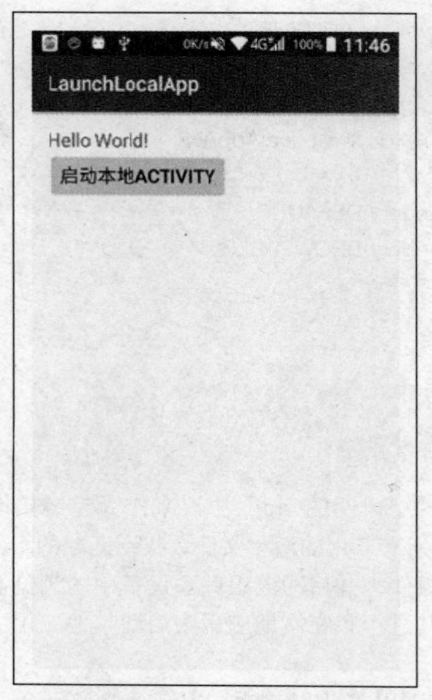

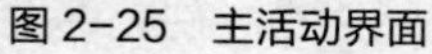
图 2-25　主活动界面

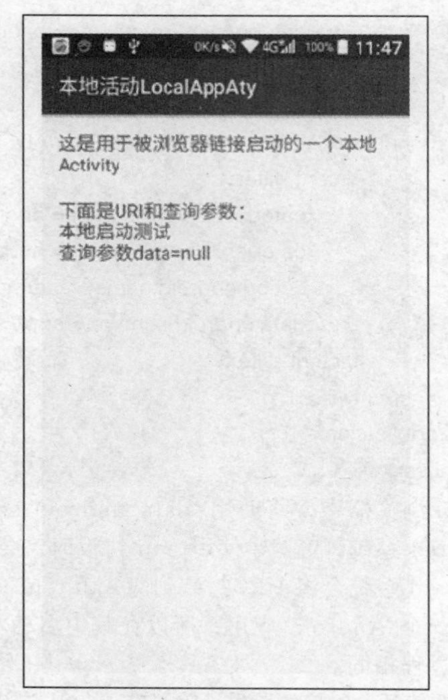

图 2-26　LocalAppAty 界面

（9）编写一个 HTML 文件，其名称为 launchaty.htm，代码如下。

```
<html>
    <style>
        a{font-size:50pt}
    </style>
<body>
    <a href="app://hello?data=Android编程">启动本地Activity</a>
</body>
</html>
```

（10）将 launchaty.htm 放在本地计算机的 Web 服务器根目录中，例如 IIS 默认 Web 服务器根目录 C:\inetpub\wwwroot。

（11）在计算机浏览器中打开 localhost/launchaty.htm，查看其运行情况，如图 2-27 所示。此时，单击页面中的超链接并不能打开本地活动。

图 2-27　在计算机浏览器中打开网页

（12）在 Android 模拟器中启动浏览器，使用计算机 IP 地址作为主机名，打开 launchaty.htm，如图 2-28 所示。单击页面中的超链接打开本地活动，如图 2-29 所示。

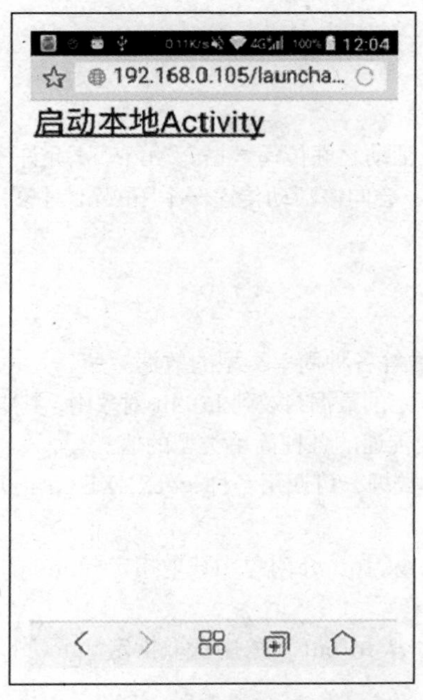

图 2-28　在手机浏览器中打开网页

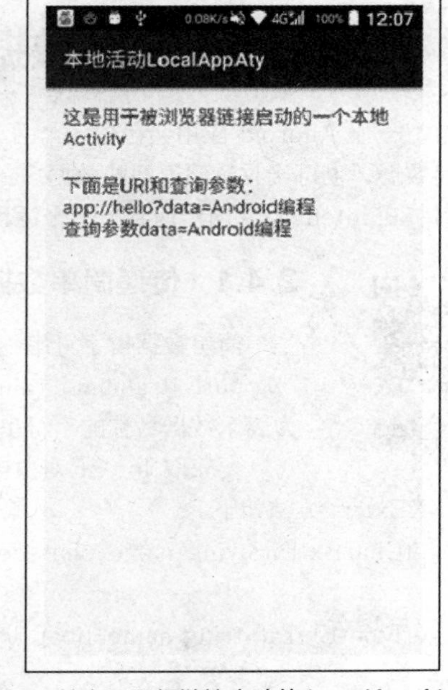

图 2-29　单击网页超链接启动的 LocalAppAty 界面

其实，在计算机中配置 Web 服务器，并编写网页来启动本地活动，这个测试的过程略微复杂。在测试设备的浏览器地址栏中直接输入以“app:/”开头的任意地址，均可启动 LocalAppAty 活动，而不需要编写网页。例如，图 2-30 显示了在浏览器地址栏中输入“app://test?data=测试”时启动的 LocalAppAty 活动界面。

图 2-30　在浏览器地址栏中输入 URL 启动活动

2.4 在活动之间传递数据

通常，一个 Android 应用会包含多个活动，也需要在活动之间传递数据。Android 允许向启动的活动传递数据，也可接收活动返回的数据。在启动活动时，会向该活动传递一个 Intent 对象，各种数据通过封装在 Intent 对象内传递到另一个活动中。

传递简单数据

2.4.1 传递简单数据

简单数据指字符串、整数、浮点数等各种简单类型的数据。

putExtra(name,value)方法可将指定的数据封装到 Intent 对象中，其中，name 为表示数据名称的字符串，value 为要传递的各种简单类型的值。

要获取 Intent 对象中封装的简单数据，可调用各种 getXXXExtra()方法。部分 getXXXExtra()方法如下。

- getCharExtra(String name,char defaultValue)：从 Intent 对象中获取指定 name 的 char 类型数据。
- getFloatExtra(String name,float defaultValue)：从 Intent 对象中获取指定 name 的 float 类型数据。
- getFloatArrayExtra(String name)：从 Intent 对象中获取指定 name 的 float 类型数组。
- getIntArrayExtra(String name)：从 Intent 对象中获取指定 name 的 int 类型数组。
- getIntExtra(String name, int defaultValue)：从 Intent 对象中获取指定 name 的 int 类型数据。
- getStringArrayExtra(String name)：从 Intent 对象中获取指定 name 的 String 类型数组。
- getStringExtra(String name)：从 Intent 对象中获取指定 name 的 String 类型数据。

各种 getXXXExtra()方法中的 defaultValue 表示默认值。如果 Intent 对象中没有指定 name 的数据，则将默认值作为方法返回值。可用 hasExtra(String name)方法判断 Intent 对象中是否包含指定 name 的数据。

下面通过实例说明如何使用 Intent 对象向活动传递简单数据，具体操作步骤如下。

（1）在 Android Studio 中创建一个新项目，将应用名称设置为 SendSimpleData，并为项目添加一个空活动。

（2）修改 activity_main.xml，在主活动布局中添加一个按钮控件，主要代码如下。

```
<?xml version="1.0" encoding="utf-8"?>
<RelativeLayout xmlns:android="http://schemas.android.com/apk/res/android"
    ...>
    <TextView android:id="@+id/textView"
        android:layout_width="wrap_content"
        android:layout_height="wrap_content"
        android:text="Hello World!" />
    <Button
        android:text="启动另一个Activity"
        android:layout_width="wrap_content"
        android:layout_height="wrap_content"
        android:layout_below="@+id/textView"
        android:id="@+id/button" />
</RelativeLayout>
```

（3）为项目添加一个空活动，活动名称为 ReceiveDataActivity。

（4）修改 activity_receive_data.xml，为布局添加一个文本视图控件，用于显示接收到的数据，主要代码如下。

```
<?xml version="1.0" encoding="utf-8"?>
<RelativeLayout xmlns:android="http://schemas.android.com/apk/res/android"
    ...>
    <TextView
        android:text="TextView"
        android:layout_width="wrap_content"
        android:layout_height="wrap_content"
        android:id="@+id/textView2" />
</RelativeLayout>
```

（5）修改 ReceiveDataActivity.java，在 onCreate()方法中获取 Intent 对象中的数据，将其显示到文本视图控件中，主要代码如下。

```
package com.example.pad.sendsimpledata;
import android.content.Intent;
...
public class ReceiveDataActivity extends AppCompatActivity {
    @Override
    protected void onCreate(Bundle savedInstanceState) {
        super.onCreate(savedInstanceState);
        setContentView(R.layout.activity_receive_data);
        Intent intent=getIntent();
        TextView tv= (TextView) findViewById(R.id.textView2);
        String name=intent.getStringExtra("name");
        int age=intent.getIntExtra("age",0);
        tv.setText("接收到的数据如下：\nname="+name+"\nage="+age);
    }
}
```

（6）修改 MainActivity.java，添加按钮的单击事件监听器，在单击按钮时，首先创建 Intent 对象，然后在其中封装数据，最后用其启动活动，主要代码如下。

```
package com.example.pad.sendsimpledata;
import android.content.Intent;
...
public class ReceiveDataActivity extends AppCompatActivity {
    @Override
    protected void onCreate(Bundle savedInstanceState) {
        super.onCreate(savedInstanceState);
        setContentView(R.layout.activity_receive_data);
        TextView tv= (TextView) findViewById(R.id.textView2);
        //从Intent对象获取简单数据
        Intent intent=getIntent();
        String name=intent.getStringExtra("name");
        int age=intent.getIntExtra("age",0);
        tv.setText("接收到的数据如下：\nname="+name+"\nage="+age);
```

```
    }
}
```

（7）运行项目，测试运行效果。

项目运行的主活动界面如图 2-31 所示，单击“启动另一个 ACTIVITY”按钮启动另一个活动，如图 2-32 所示，图中显示了接收到的数据。

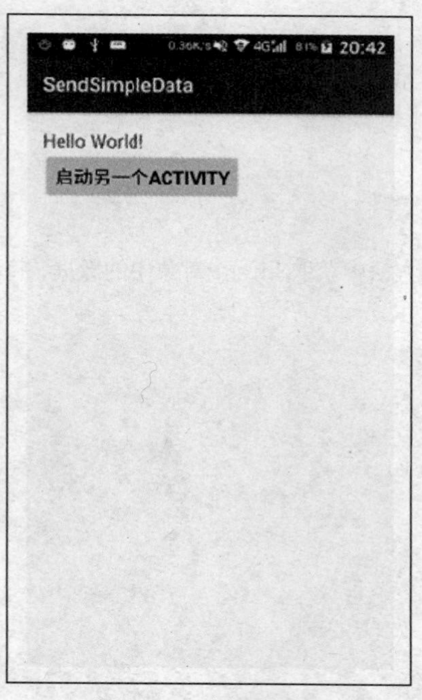

图 2-31　主活动界面

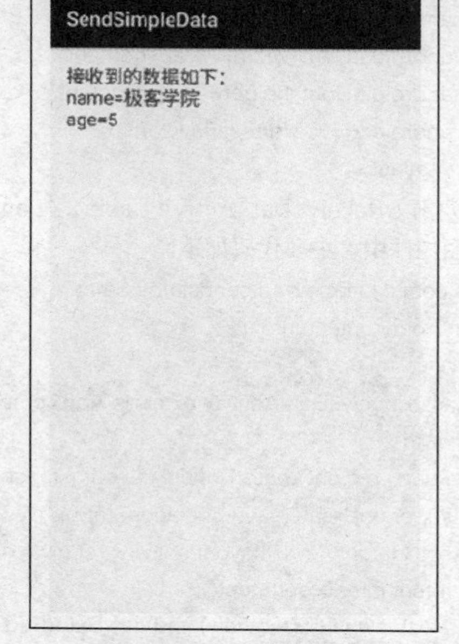

图 2-32　显示接收到的数据

传递数据包 Bundle

2.4.2　传递 Bundle 对象

Bundle 对象可用于封装各种简单数据，再将其封装到 Intent 对象中传递给活动。

Bundle 对象的各种 putXXX(String key,XXX value)方法，可将 XXX 类型的数据封装到其中，对应的用 getXXX(String key)方法可从其中获取数据，Bundle 对象的用法和 Intent 对象类似。之所以不直接使用 Intent 对象来封装数据，是为了用 Bundle 对象来对数据打包，这在大型项目协作开发过程中非常适用。

Bundle 对象准备好之后，调用 putExtras(bundle)或 putExtra(name,bundle)方法可将其封装到 Intent 对象中。要从 Intent 对象中获取 Bundle 对象时，调用对应的 getExtras()或 getBundleExtra()方法即可。

这里使用 2.4.1 小节中创建的实例项目 SendSimpleData 来测试如何向活动传递 Bundle 对象，具体操作步骤如下。

（1）修改 MainActivity.java，将数据封装到 Bundle 对象之后，再将 Bundle 对象封装到 Intent 对象中，最后用其启动活动，主要代码如下。

```
package com.example.pad.sendsimpledata;
import android.content.Intent;
...
public class MainActivity extends AppCompatActivity {
```

```
    @Override
    protected void onCreate(Bundle savedInstanceState) {
        super.onCreate(savedInstanceState);
        setContentView(R.layout.activity_main);
        findViewById(R.id.button).setOnClickListener(new View.OnClickListener() {
            @Override
            public void onClick(View v) {
                Intent intent=new Intent(MainActivity.this,ReceiveDataActivity.class);
                //传递简单数据
                //intent.putExtra("name","极客学院");
                //intent.putExtra("age",5);

                //传递Bundle对象
                Bundle bd=new Bundle();
                bd.putString("name","极客学院");
                bd.putInt("age",5);
                intent.putExtras(bd);
                startActivity(intent);
            }
        });
    }
}
```

（2）修改 ReceiveDataActivity.java，首先从 Intent 对象中获取 Bundle 对象，再从其中获取数据，并将其显示到文本视图控件中，代码如下。

```
package com.example.pad.sendsimpledata;
import android.content.Intent;
...
public class ReceiveDataActivity extends AppCompatActivity {
    @Override
    protected void onCreate(Bundle savedInstanceState) {
        super.onCreate(savedInstanceState);
        setContentView(R.layout.activity_receive_data);
        TextView tv= (TextView) findViewById(R.id.textView2);
        //从Intent对象获取简单数据
        Intent intent=getIntent();
        //String name=intent.getStringExtra("name");
        //int age=intent.getIntExtra("age",0);
        //从Intent对象获取Bundle对象
        Bundle bd=intent.getExtras();
        String name=bd.getString("name");
        int age=bd.getInt("age");
        tv.setText("接收到的数据如下：\nname="+name+"\nage="+age);
    }
}
```

修改后的项目运行结果和 2.4.1 小节中的完全相同。

传递对象

2.4.3 传递对象

如果需要在活动间传递自定义的类对象，又该如何做呢？

自定义的类对象不能像简单数据一样直接封装到 Intent 对象中，Android 系统要求封装到 Intent 对象中，支持序列化。让类实现 Java 内置的 Serializable 接口，或者实现 Android 提供的 Parcelable 接口，即可使类对象支持序列化。

1. 使用实现 Serializable 接口的类对象

下面通过实例说明如何向活动传递实现了 Serializable 接口的类对象，具体操作步骤如下。

（1）在 Android Studio 中创建一个新项目，将应用名称设置为 SendObject，并为项目添加一个空活动。

（2）修改 activity_main.xml，在主活动布局中添加一个按钮控件，代码如下。

```
<?xml version="1.0" encoding="utf-8"?>
<RelativeLayout xmlns:android="http://schemas.android.com/apk/res/android"
    ...>
    <Button
        android:text="启动另一个Activity"
        android:layout_width="wrap_content"
        android:layout_height="wrap_content"
        android:id="@+id/button" />
</RelativeLayout>
```

（3）为项目添加一个 Java 类，类名称为 User，代码如下。

```
package com.example.xbg.sendobject;
import java.io.Serializable;
public class User implements Serializable {
    private String name;
    private int age;
    public User(String name, int age) {
        this.name = name;
        this.age = age;
    }
    public String getName() { return name; }
    public void setName(String name) {this.name = name; }
    public int getAge() { return age; }
    public void setAge(int age) { this.age = age; }
}
```

（4）为项目添加一个空活动，活动名称为 ReceiveDataActivity。

（5）修改 activity_receive_data.xml，为布局添加一个文本视图控件，用于显示接收到的数据，代码如下。

```
<?xml version="1.0" encoding="utf-8"?>
<RelativeLayout xmlns:android="http://schemas.android.com/apk/res/android"
    ...>
    <TextView
        android:text="TextView"
        android:layout_width="wrap_content"
```

```
        android:layout_height="wrap_content"
        android:id="@+id/textView2" />
</RelativeLayout>
```

（6）修改 MainActivity.java，添加按钮的单击事件监听器，在单击按钮时，首先创建 Intent 对象，然后在其中封装 User 对象，最后用其启动活动，代码如下。

```
package com.example.xbg.sendobject;
import android.content.Intent;
...
public class MainActivity extends AppCompatActivity {
    @Override
    protected void onCreate(Bundle savedInstanceState) {
        super.onCreate(savedInstanceState);
        setContentView(R.layout.activity_main);
        findViewById(R.id.button).setOnClickListener(new View.OnClickListener() {
            @Override
            public void onClick(View v) {
                Intent intent=new Intent(MainActivity.this,ReceiveDataActivity.class);
                intent.putExtra("user",new User("极客学院",5));
                startActivity(intent);
            }
        });
    }
}
```

（7）修改 ReceiveDataActivity.java，在 onCreate()方法中获取 Intent 对象中的数据，并将其显示到文本视图控件中，代码如下。

```
package com.example.xbg.sendobject;
import android.content.Intent;
...
public class ReceiveDataActivity extends AppCompatActivity {
    @Override
    protected void onCreate(Bundle savedInstanceState) {
        super.onCreate(savedInstanceState);
        setContentView(R.layout.activity_receive_data);
        Intent intent=getIntent();
        User user= (User) intent.getSerializableExtra("user");
        TextView tv= (TextView) findViewById(R.id.textView2);
        String name=user.getName();
        int age=user.getAge();
        tv.setText("接收到的User对象：User(name:"+name+",age:"+age+")");
    }
}
```

（8）运行项目，测试运行效果。

项目运行时，主活动界面如图 2-33 所示，单击“启动另一个 ACTIVITY”按钮可切换到显示接收数据的活动界面，如图 2-34 所示。

图 2-33　主活动界面

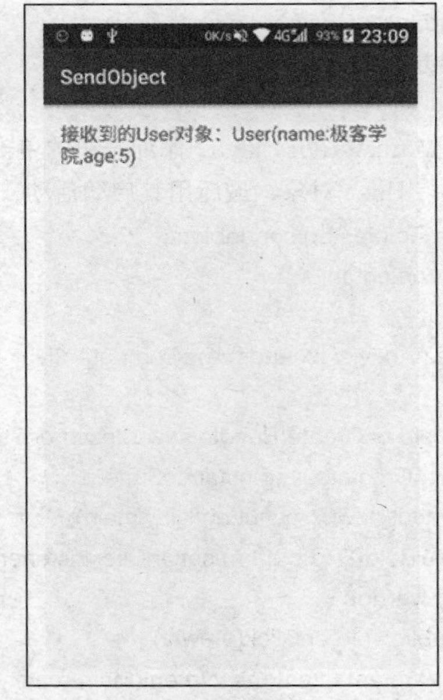

图 2-34　接收数据的活动界面

2. 使用实现 Parcelable 接口的类对象

与实现 Serializable 接口相比，实现 Parcelable 接口需要编写更多代码。

如下实例将通过修改前文创建的 SendObject 项目来说明如何向活动传递实现了 Parcelable 接口的类对象，具体操作步骤如下。

（1）修改 User.java，实现 Parcelable 接口，代码如下。

```
package com.example.xbg.sendobject;
import android.os.Parcel;
import android.os.Parcelable;
public class User implements Parcelable {
    private String name;
    private int age;
    public User(String name, int age) {
        this.name = name;
        this.age = age;
    }
    public String getName() { return name; }
    public void setName(String name) { this.name = name; }
    public int getAge() { return age; }
    public void setAge(int age) { this.age = age; }
    @Override
    public int describeContents() {
        return 0;
    }
    @Override
    public void writeToParcel(Parcel dest, int flags) {
        //将对象数据成员写入dest
```

```
            dest.writeString(getName());
            dest.writeInt(getAge());
        }
        public static  Creator<User> CREATOR=new Creator<User>() {
            @Override
            public User createFromParcel(Parcel source) {
                return new User(source.readString(),source.readInt());
            }
            @Override
            public User[] newArray(int size) {
                return new User[size];
            }
        };
}
```

（2）修改 ReceiveDataActivity.java，用 getParcelableExtra()方法来获取 Intent 中的 User 对象，代码如下。

```
package com.example.xbg.sendobject;
import android.content.Intent;
...
public class ReceiveDataActivity extends AppCompatActivity {
    @Override
    protected void onCreate(Bundle savedInstanceState) {
        super.onCreate(savedInstanceState);
        setContentView(R.layout.activity_receive_data);
        Intent intent=getIntent();
        //User user= (User) intent.getSerializableExtra("user");        //类实现Serializable接口时用
        User user= (User) intent.getParcelableExtra("user");          //类实现Parcelable接口时用
        TextView tv= (TextView) findViewById(R.id.textView2);
        String name=user.getName();
        int age=user.getAge();
        tv.setText("接收到的User对象：User(name:"+name+",age:"+age+")");
    }
}
```

（3）运行项目，可看到修改后的项目运行效果与修改前没有区别。

2.4.4 获取活动返回的数据

获取 Activity 的返回数据

通过项目设置，不仅可以向启动的活动传递数据，也可获取活动中返回的数据。

前文多个实例中使用了 startActivity()方法来启动一个活动，并在 Intent 对象中封装需要向活动传递的数据。要获得活动中返回的数据，则需要使用 startActivityForResult(intent,requestCode)方法来启动活动。其中，参数 intent 是一个 Intent 对象，用于封装需要传递给活动的数据。参数 requestCode 为请求码，是一个整数，用来标识当前请求。一个活动可能会接收到其他不同活动的请求，从活动返回时，它会原样返回接收到的请求码。在处理返回结果时，可通过请求码判断是不是从所请求的活动返回。

当前活动中需重载 onActivityResult()方法来处理返回结果，其代码基本结构如下。

```
protected void onActivityResult(int requestCode, int resultCode, Intent data) {
        super.onActivityResult(requestCode, resultCode, data);
...
}
```

参数 requestCode 为从所请求的活动返回的它所接收到的请求码；resultCode 为结果代码，常量 RESULT_CANCELED 表示用户取消了操作，RESULT_OK 表示用户正确完成了操作；data 为请求活动返回的 Intent 对象，从中可获取返回的数据。

在请求的活动中，用 setResult(resultCode,intent)方法设置返回结果，resultCode 为结果代码，intent 为封装了返回数据的 Intent 对象。

下面通过实例说明如何启动一个活动并获取从活动返回的数据，具体操作步骤如下。

（1）创建一个新项目，将应用名称设置为 ReceiveDataFromActivity，并为项目添加一个空活动。

（2）修改 activity_main.xml，在主活动布局中添加一个按钮控件，代码如下。

```
<?xml version="1.0" encoding="utf-8"?>
<RelativeLayout xmlns:android="http://schemas.android.com/apk/res/android"
        ...>
    <Button
        android:text="启动另一个Activity"
        android:layout_width="wrap_content"
        android:layout_height="wrap_content"
        android:id="@+id/button" />
    <TextView   android:id="@+id/textView"
        android:layout_width="wrap_content"
        android:layout_height="wrap_content"
        android:layout_below="@id/button"
        android:text="Hello World!" />
</RelativeLayout>
```

（3）为项目添加一个空活动，活动名称为 SendDataBack。

（4）修改 activity_send_data_back.xml，为布局添加一个编辑视图控件和一个按钮控件，代码如下。

```
<?xml version="1.0" encoding="utf-8"?>
<LinearLayout xmlns:android="http://schemas.android.com/apk/res/android"
        ...>
    <EditText
        android:layout_width="match_parent"
        android:layout_height="wrap_content"
        android:text="输入数据"
        android:id="@+id/editText" />
    <Button
        android:text="确认返回"
        android:layout_width="match_parent"
        android:layout_height="wrap_content"
        android:id="@+id/button2" />
</LinearLayout>
```

（5）修改 SendDataBack.java，添加按钮单击事件监听器，在单击按钮时，将输入的数据封装到

Intent 对象中，返回给上一个活动，代码如下。

```
package com.example.xbg.receivedatafromactivity;
import android.content.Intent;
...
public class SendDataBack extends AppCompatActivity {
    @Override
    protected void onCreate(Bundle savedInstanceState) {
        super.onCreate(savedInstanceState);
        setContentView(R.layout.activity_send_data_back);
        findViewById(R.id.button2).setOnClickListener(new View.OnClickListener() {
            @Override
            public void onClick(View v) {
                Intent intent=new Intent();
                EditText editText= (EditText) findViewById(R.id.editText);
                intent.putExtra("data",editText.getText().toString());//将输入装入Intent
                setResult(RESULT_OK,intent);//设置返回结果
                finish();//结束当前活动
            }
        });
    }
}
```

（6）修改 MainActivity.java，添加按钮的单击事件监听器，在单击按钮时启动另一个活动，同时还需要处理返回结果，代码如下。

```
package com.example.xbg.receivedatafromactivity;
import android.content.Intent;
...
public class MainActivity extends AppCompatActivity {
    private   static int REQUEST_CODE=1011;//标识当前活动的请求
    @Override
    protected void onCreate(Bundle savedInstanceState) {
        super.onCreate(savedInstanceState);
        setContentView(R.layout.activity_main);
        findViewById(R.id.button).setOnClickListener(new View.OnClickListener() {
            @Override
            public void onClick(View v) {
                Intent intent=new Intent(MainActivity.this,SendDataBack.class);
                startActivityForResult(intent,REQUEST_CODE);//启动可返回结果的活动
            }
        });
    }
    @Override
    protected void onActivityResult(int requestCode, int resultCode, Intent data) {
        super.onActivityResult(requestCode, resultCode, data);
        if(requestCode==REQUEST_CODE){//返回的请求码与当前活动请求码一致时，才执行后继操作
            if(resultCode==RESULT_OK){//RESULT_OK表示返回的活动已成功处理请求
```

```
                TextView tv= (TextView) findViewById(R.id.textView);
                //将另一个活动返回的Intent中的数据显示到文本视图中
                tv.setText("从另一个Activity返回的数据是："+data.getStringExtra("data"));
            }
        }
}}
```

（7）运行项目，测试运行效果。

项目运行时，主活动界面如图 2-35 所示。单击“启动另一个 ACTIVITY”按钮启动另一个活动，界面如图 2-36 所示。将编辑视图控件中的文字修改为“极客”，再单击“确认返回”按钮返回主活动界面，主活动显示返回的数据，界面如图 2-37 所示。

图2-35　主活动

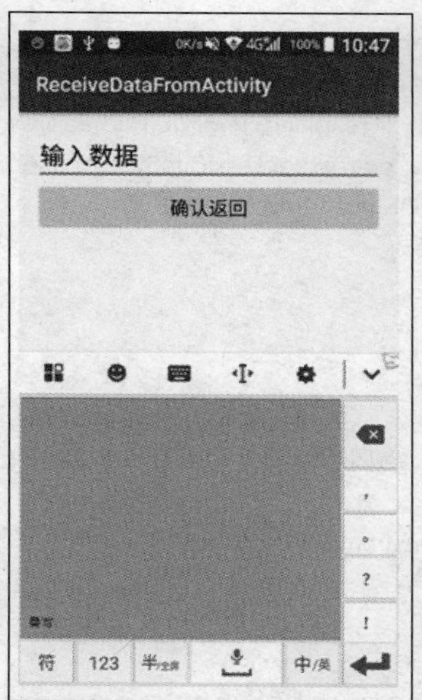

图2-36　另一个活动界面

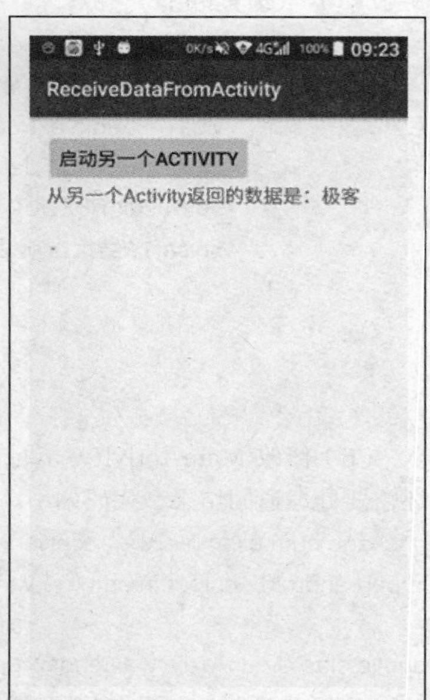

图2-37　显示返回数据的主活动界面

2.5　活动的生命周期

活动的生命周期指活动从第一次创建到被销毁的整个时间。在一个生命周期内，活动可能存在多种状态。深入了解活动的生命周期，有助于开发人员更合理地管理应用程序资源，设计出效率更高的应用。

查看活动帮助文档

2.5.1　返回栈、活动状态及生命周期回调

通常，一个应用可能包含多个活动。Android 系统使用堆栈（也称返回栈）来管理活动。

一个活动可启动另一个活动，甚至是其他应用的活动。Android 将运行一个应用时间打开的所有活动（不管是否属于当前应用）统称为一个任务。同一个任务中打开的活动将按照先后顺序排列在堆栈中。Android 通过堆栈统一管理任务中各个活动的状态。

1. 返回栈

返回栈遵循“先进后出”的原则。通常，设备的主屏幕是大多数任务的起点。当用户在屏幕上触摸应用图标或应用快捷方式时，该应用被启动并出现在前台。如果应用没有被启动过，系统就为应用创建一个新任务，应用的“主”活动将作为任务返回栈的第一个活动（也称根活动）。

当前活动启动另一个活动时，新活动会放到堆栈顶部，成为焦点所在，即新的当前活动。前一个活动仍保留在堆栈中，只是处于停止状态。活动停止时，系统会为其保存用户界面状态。当用户按返回键时，当前活动从堆栈顶部弹出（被销毁），它下面的一个活动恢复执行（恢复其入栈前的状态）。用户能访问的活动永远都位于堆栈顶部，只有当栈顶活动被弹出后，它下面的一个活动才能成为当前活动。Android 不会改变堆栈中活动的顺序，只有活动入栈和出栈两种操作。

图 2-38 所示为任务返回栈的活动情况，具有下列默认行为特点。

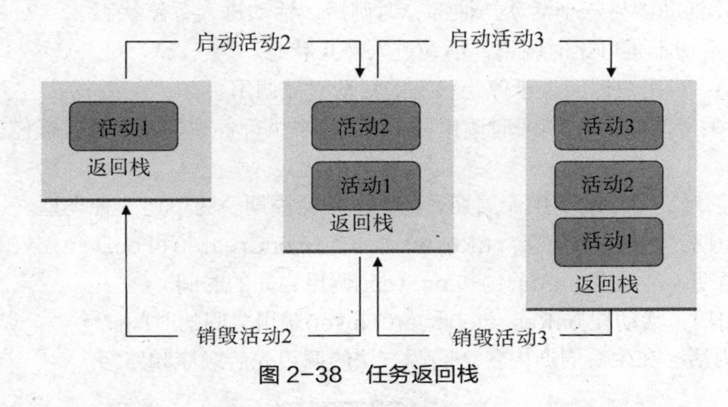

图 2-38　任务返回栈

- 当活动 1 启动活动 2 时，活动 1 停止，系统会保持其状态（例如用户输入会保留）。当在活动 2 中按返回键时，活动 1 恢复执行。
- 从当前活动按返回键时，当前活动从任务栈顶部弹出并销毁。
- 当按主页键返回设备主屏幕时，当前任务进入后台，系统保留任务中的每个活动。当用户通过应用图标或概览屏幕返回任务时，系统将恢复任务，并恢复任务返回栈顶部的活动状态。

一种例外的情况是，当用户离开任务的时间过长，系统默认只为任务保留根活动，其他活动都会被销毁。当用户再次返回任务时，系统只能恢复根活动。可在清单文件中通过如下属性设置告诉系统在用户长时间离开任务时如何处理活动。

- alwaysRetainTaskState：如果在任务的根活动中将此属性设置为 true，则在用户离开很长一段时间后，系统仍会保留任务的所有活动，即任务返回栈保持不变。
- clearTaskOnLaunch：如果在任务的根活动中将此属性设置为 true，则在用户离开很长一段时间后，系统只为任务保留根活动。
- finishOnTaskLaunch：此属性适用于单个活动，设置为 true 时，只要用户离开任务，活动就会被销毁。

2. 活动状态

活动在其生存周期内可能有下述 4 种状态。

- 运行状态：活动位于任务返回栈顶部时，活动就处于运行状态。用户只能和运行状态的活动交互。
- 暂停状态：当另一个活动位于屏幕最前面，但没有覆盖整个屏幕，活动仍然部分可见时，活动进入暂停状态。系统会保持暂停状态活动的全部状态，除非系统内存极度缺乏，否则系统不会销毁暂停状态的活动。
- 停止状态：当活动被另一个活动完全覆盖时，活动进入停止状态。系统会保持停止状态活动的全部状态，但可能会随时销毁活动。
- 销毁状态：系统会从内存删除被销毁（也称被回收）活动的所有资源。用户通过导航再次返回该活动时，系统会重建该活动。

3. 活动的生命周期回调

当活动在各种状态之间转换时，系统会执行各种回调方法，开发人员可在各种回调方法中编写代码实现各种操作。如下所述是活动的各个生命周期回调方法。

- onCreate()：活动正在被创建时调用。通常在该方法中完成各种初始化操作，例如绑定布局、定义监听器等。
- onStart()：活动正在被显示到屏幕最前面时调用，活动即将进入运行状态。
- onResume()：活动已经被显示到屏幕最前面时调用，活动已进入运行状态，准备好与用户交互。
- onPause()：活动被另一个活动挡住部分时调用，活动进入暂停状态。
- onStop()：活动不再可见时调用，活动进入停止状态。
- onRestart()：当用户返回处于停止状态的活动时被调用。
- onDestroy()：活动正在被销毁时调用，通常在其中执行各种资源的回收操作。方法执行结束时，活动从内存中删除。

所有这些方法都定义了活动的生命周期，活动的生命周期分为下述 3 种类型。

- 完整生命周期：活动经历了除 onRestart()外，在 onCreate()和 onDestroy()调用之间的时间。
- 可见生命周期：活动在 onStart()和 onStop()调用之间的时间。
- 前台生命周期：活动在 onResume()和 onPause()调用之间的时间。

图 2-39 所示为活动在生命周期内各个回调方法的调用先后顺序和路径。

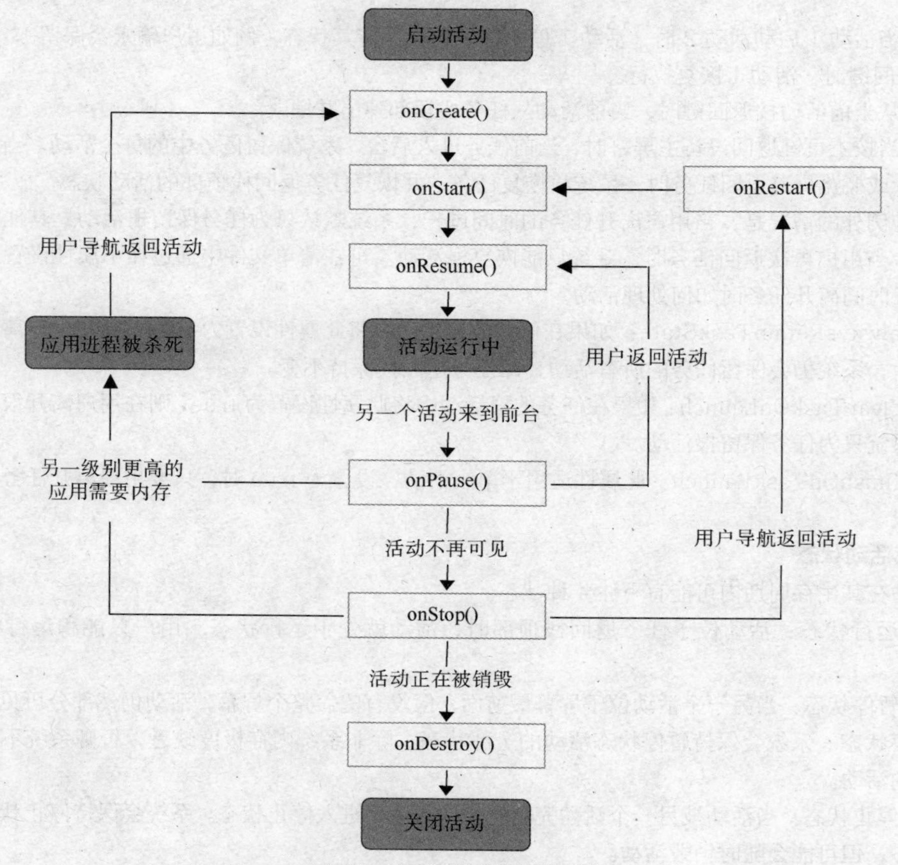

图 2-39　活动生命周期

2.5.2 检验活动的生命周期

认识 Activity 生命周期

下面通过一个实例来检验活动的生命周期，具体操作步骤如下。

（1）在 Android Studio 中创建一个新项目，将应用名称设置为 ActivityLifeCircle，并为项目添加一个空活动。

（2）修改 activity_main.xml，在主活动布局中添加一个按钮控件，代码如下。

```
<?xml version="1.0" encoding="utf-8"?>
<RelativeLayout xmlns:android="http://schemas.android.com/apk/res/android"
        ...>
    <TextView android:id="@+id/textView"
        android:layout_width="wrap_content"
        android:layout_height="wrap_content"
        android:text="Hello World!" />
    <Button
        android:text="启动另一个活动"
        android:layout_width="wrap_content"
        android:layout_height="wrap_content"
        android:layout_below="@+id/textView"
        android:id="@+id/button" />
</RelativeLayout>
```

（3）为项目添加一个空活动，活动名称为 Main2Activity。

（4）修改 Main2Activity.java，重载主活动的各个生命周期回调方法，代码如下。

```
package com.example.xbg.activitylifecircle;
import android.support.v7.app.AppCompatActivity;
import android.os.Bundle;
public class Main2Activity extends AppCompatActivity {
    @Override
    protected void onCreate(Bundle savedInstanceState) {
        super.onCreate(savedInstanceState);
        setContentView(R.layout.activity_main2);
        System.out.println("正在执行Main2Activity.onCreate()");
    }
    @Override
    protected void onDestroy() {
        super.onDestroy();
        System.out.println("正在执行Main2Activity.onDestroy()");
    }
    @Override
    protected void onStart() {
        super.onStart();
        System.out.println("正在执行Main2Activity.onStart()");
    }
    @Override
    protected void onStop() {
        super.onStop();
```

```
        System.out.println("正在执行Main2Activity.onStop()");
    }
    @Override
    protected void onPause() {
        super.onPause();
        System.out.println("正在执行Main2Activity.onPause()");
    }
    @Override
    protected void onResume() {
        super.onResume();
        System.out.println("正在执行Main2Activity.onResume()");
    }
    @Override
    protected void onRestart() {
        super.onRestart();
        System.out.println("正在执行Main2Activity.onRestart()");
    }
}
```

（5）修改 MainActivity.java，添加按钮的单击事件监听器，在单击按钮时启动活动 Main2Activity，同时重载主活动的各个生命周期回调方法，代码如下。

```
package com.example.xbg.activitylifecircle;
import android.content.Intent;
...
public class MainActivity extends AppCompatActivity {
    @Override
    protected void onCreate(Bundle savedInstanceState) {
        super.onCreate(savedInstanceState);
        setContentView(R.layout.activity_main);
        System.out.println("正在执行MainActivity.onCreate()");
        findViewById(R.id.button).setOnClickListener(new View.OnClickListener() {
            @Override
            public void onClick(View v) {
                Intent intent=new Intent(MainActivity.this,Main2Activity.class);
                startActivity(intent);
            }
        });
    }
    @Override
    protected void onDestroy() {
        super.onDestroy();
        System.out.println("正在执行MainActivity.onDestroy()");
    }
    @Override
    protected void onStart() {
        super.onStart();
```

```
        System.out.println("正在执行MainActivity.onStart()");
    }
    @Override
    protected void onStop() {
        super.onStop();
        System.out.println("正在执行MainActivity.onStop()");
    }
    @Override
    protected void onPause() {
        super.onPause();
        System.out.println("正在执行MainActivity.onPause()");
    }
    @Override
    protected void onResume() {
        super.onResume();
        System.out.println("正在执行MainActivity.onResume()");
    }
    @Override
    protected void onRestart() {
        super.onRestart();
        System.out.println("正在执行MainActivity.onRestart()");
    }
}
```

（6）运行项目，测试运行效果，主活动界面如图 2-40 所示。

图 2-40 主活动界面

在 Android Studio 的 Run 窗口中可看到系统执行了主活动 MainActivity 的生命周期回调方法的

输出信息，如图 2-41 所示。在应用启动并显示主活动的过程中，执行了主活动的 onCreate()、onStart() 和 onResume()等生命周期回调方法。

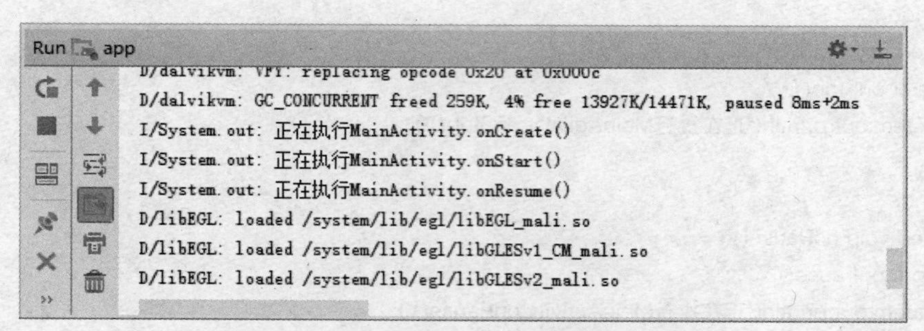

图 2-41　启动活动后查看执行的回调方法

单击 Run 窗口左侧边栏中的 按钮，可清除窗口中当前显示的输出信息，以便更清楚地了解在执行当前操作时应用在后台的输出信息。

按返回键返回主屏幕，查看 Run 窗口信息，如图 2-42 所示，可看到此过程中系统执行了主活动的 onPause()、onStop()和 onDestroy()等生命周期回调方法。

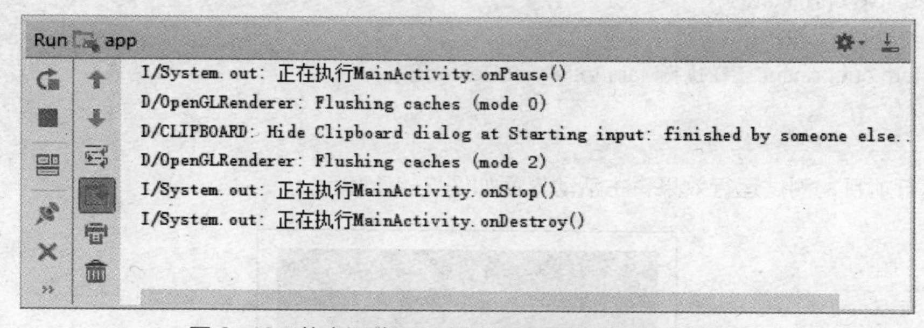

图 2-42　从主活动返回主屏幕后查看执行的回调方法

从设备概览屏幕返回应用，查看 Run 窗口信息，如图 2-43 所示，可看到此过程中系统执行了主活动的 onCreate()、onStart()和 onResume()等生命周期回调方法，与第一次启动应用时相同。

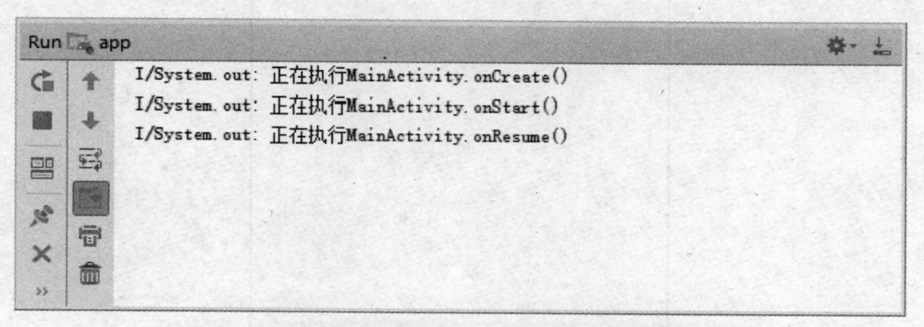

图 2-43　从概览屏幕返回应用后查看执行的回调方法

在主活动界面中单击“启动另一个活动”按钮，启动活动 Main2Activity，查看 Run 窗口信息，如图 2-44 所示，可看到此过程中系统按顺序执行的生命周期回调方法包括主活动的 onPause()，活动 Main2Activity 的 onCreate()、onStart()和 onResume()，主活动的 onStop()和 onDestroy()等。

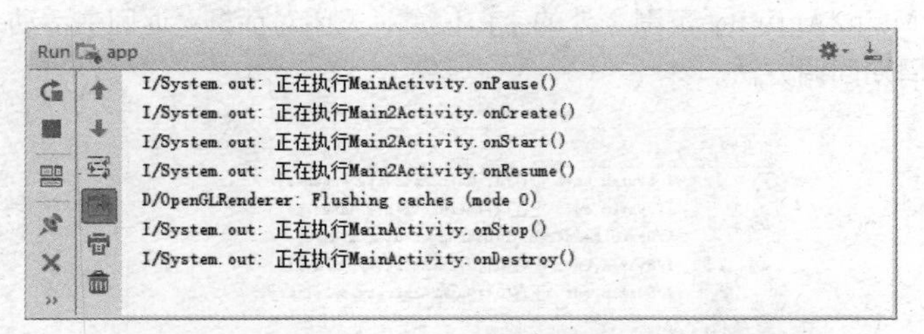

图 2-44　从主活动启动另一个活动后查看执行的回调方法

按返回键返回主活动，查看 Run 窗口信息，如图 2-45 所示，可看到此过程中系统按顺序执行的生命周期回调方法包括活动 Main2Activity 的 onPause()，主活动的 onCreate()、onStart()和 onResume()，活动 Main2Activity 的 onStop()和 onDestroy()等。

```
Run app
I/System.out: 正在执行Main2Activity.onPause()
I/System.out: 正在执行MainActivity.onCreate()
I/System.out: 正在执行MainActivity.onStart()
I/System.out: 正在执行MainActivity.onResume()
D/OpenGLRenderer: Flushing caches (mode 0)
I/System.out: 正在执行Main2Activity.onStop()
I/System.out: 正在执行Main2Activity.onDestroy()
```

图 2-45　从另一个活动返回活动后查看执行的回调方法

如果是从活动 Main2Activity 直接返回设备主屏幕，则系统输出如图 2-46 所示，可看到此过程中系统按顺序执行了活动 Main2Activity 的 onPause()、onStop()和 onDestroy()等生命周期回调方法。

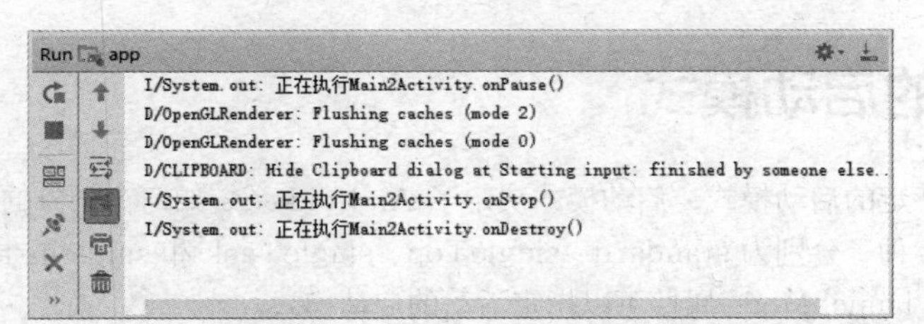

图 2-46　从活动 Main2Activity 直接返回设备主屏幕后查看执行的回调方法

修改清单文件 AndroidManifest.xml，为活动 Main2Activity 添加主题，以对话框的样式显示，代码如下。

```
<activity android:name=".Main2Activity"
        android:label="@string/app_name"
        android:theme="@style/Base.Theme.AppCompat.Dialog"></activity>
```

修改后运行项目，从主活动中单击“启动另一个活动”按钮启动另一个活动 Main2Activity，活动 Main2Activity 以对话框样式显示，主活动只是进入了暂停状态。图 2-47 所示为此过程中执行的生命周期回调方法。

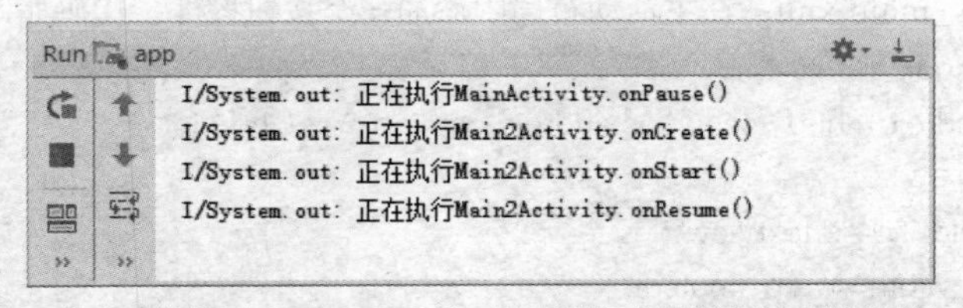

图 2-47　修改后从活动 Main2Activity 返回设备主屏幕后查看执行的回调方法

此时单击活动 Main2Activity 范围之外的空白位置，关闭对话框。返回主活动。图 2-48 所示为此过程中执行的生命周期回调方法。

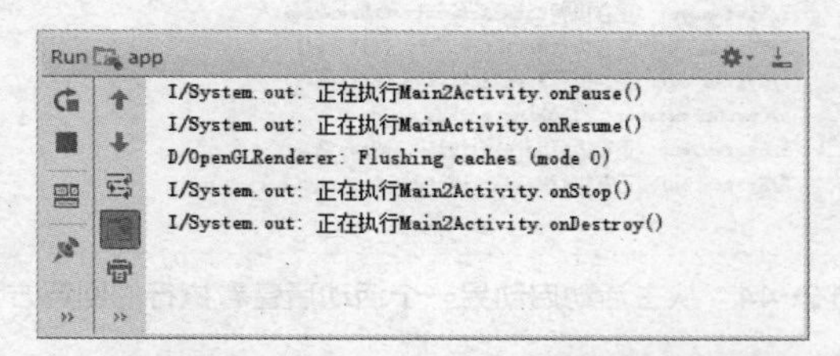

图 2-48　关闭对话框返回主活动后查看执行的回调方法

另外，在应用运行期间，如果设备的配置发生了变化（例如改变屏幕方向，更改语言或输入法等），系统会销毁并创建活动。在运行 ActivityLifeCircle 项目，显示主活动 MainActivity 时，旋转设备屏幕，Run 窗口中的输出信息如图 2-49 所示。从图中的输出可以看出，在旋转屏幕时，活动依次从运行状态进入暂停状态、停止状态、销毁状态，直到重建后重新进入运行状态。

```
Run app
D/OpenGLRenderer: Flushing caches (mode 0)
I/System.out: 正在执行MainActivity.onPause()
I/System.out: 正在执行MainActivity.onStop()
I/System.out: 正在执行MainActivity.onDestroy()
I/System.out: 正在执行MainActivity.onCreate()
I/System.out: 正在执行MainActivity.onStart()
I/System.out: 正在执行MainActivity.onResume()
```

图 2-49　运行时设备配置改变导致重建活动时执行的回调方法

2.6　活动的启动模式

活动总是拥有特定的启动模式，启动模式决定了 Android 系统如何在任务的返回栈中管理活动。活动的启动模式有 4 种，分别为 standard、singleTop、singleTask 和 singleInstance。在清单文件使用<activity>元素的 launchMode 属性可以指定活动的启动模式。

standard 模式

2.6.1　standard 模式

standard 是活动的默认启动模式。前文用到的所有活动，启动模式都是 standard。系统在启动 standard 模式的活动时，不会检查任务返回栈顶部是否已经有该活动，总是创建一个新的活动实例，并将其放到返回栈顶部。

如下实例将通过启动同一个活动来说明活动的 standard 模式，具体操作步骤如下。

（1）在 Android Studio 中创建一个新项目，将应用名称设置为 LaunchStandard，并为项目添加一个空活动。

（2）修改 activity_main.xml，在主活动布局中添加一个按钮控件，代码如下。

```
<?xml version="1.0" encoding="utf-8"?>
<LinearLayout xmlns:android="http://schemas.android.com/apk/res/android"
…>
    <TextView android:id="@+id/textView"
        android:layout_width="wrap_content"
        android:layout_height="wrap_content"
```

```
        android:text="Hello World!" />
    <Button
        android:text="启动MainActivity"
        android:layout_width="wrap_content"
        android:layout_height="wrap_content"
        android:id="@+id/button" />
</LinearLayout>
```

（3）修改 MainActivity.java，在文本视图控件中显示任务 ID 和活动实例信息，同时添加按钮的单击事件监听器，在单击按钮时启动活动 MainActivity，代码如下。

```
package com.example.xbg.launchmodel;
import android.content.Intent;
…
public class MainActivity extends AppCompatActivity {
    @Override
    protected void onCreate(Bundle savedInstanceState) {
        super.onCreate(savedInstanceState);
        setContentView(R.layout.activity_main);
        TextView tv= (TextView) findViewById(R.id.textView);
        tv.setText(String.format("任务ID：%d\n活动实例：%s",getTaskId(),this.toString()));
        findViewById(R.id.button).setOnClickListener(new View.OnClickListener() {
            @Override
            public void onClick(View v) {
                Intent intent=new Intent(MainActivity.this,MainActivity.class);
                startActivity(intent);
            }
        });
    }
}
```

（4）运行项目，测试运行效果。

项目运行时，主活动界面如图 2-50 所示，单击启动MAINACTIVITY按钮即可启动 MainActivity 活动，界面如图 2-51 所示。

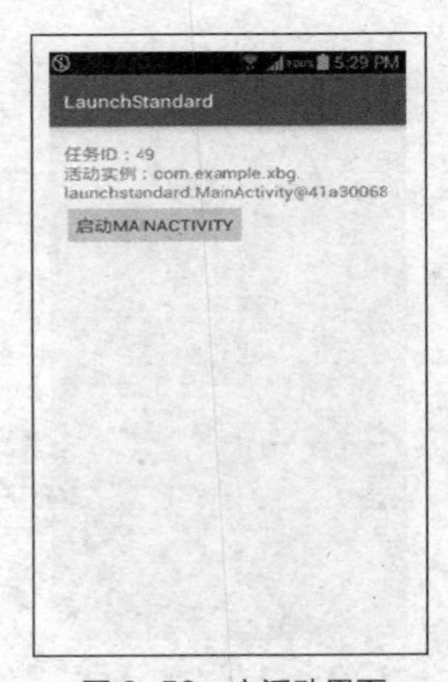

图 2-50　主活动界面

图 2-51　新的 MainActivity 活动界面

从图中可以看到，两个 MainActivity 活动实例的任务 ID 相同，说明是在同一个任务返回栈中。从两个 MainActivity 活动实例的序列化字符串可看出，两个实例是同一个类，不是同一字实例。这说明在默认的 standard 模式下，在任务中启动同一个活动时，系统会创建新的活动实例。

在第二个 MainActivity 活动中，按返回键可返回主活动界面，再按返回键才能返回设备主屏幕。

2.6.2 singleTop 模式

singleTop 模式

如果活动是 singleTop 模式，在启动活动时，系统首先检查任务返回栈。若栈顶活动是相同活动的实例，则直接使用该活动，不会再创建新的实例。

下面通过实例说明如何使用 singleTop 模式，具体操作步骤如下。

（1）在 Android Studio 中创建一个新项目，将应用名称设置为 Launch SingleTop，并为项目添加一个空活动。

（2）修改 activity_main.xml，代码如下。

```
<?xml version="1.0" encoding="utf-8"?>
<LinearLayout xmlns:android="http://schemas.android.com/apk/res/android"
    …>
    <TextView android:id="@+id/textView"
        android:layout_width="wrap_content"
        android:layout_height="wrap_content"
        android:text="Hello World!" />
    <Button
        android:text="启动MainActivity"
        android:layout_width="wrap_content"
        android:layout_height="wrap_content"
        android:id="@+id/button1" />
    <Button
        android:text="启动BActivity"
        android:layout_width="wrap_content"
        android:layout_height="wrap_content"
        android:id="@+id/button2" />
</LinearLayout>
```

（3）为项目添加一个空活动，活动名称为 BActivity。

（4）修改 activity_b.xml，代码如下。

```
<?xml version="1.0" encoding="utf-8"?>
<LinearLayout xmlns:android="http://schemas.android.com/apk/res/android"
    …>
    <TextView android:id="@+id/textView"
        android:layout_width="wrap_content"
        android:layout_height="wrap_content"
        android:text="Hello World!" />
    <Button
        android:text="启动MainActivity"
        android:layout_width="wrap_content"
        android:layout_height="wrap_content"
        android:id="@+id/button1" />
```

```
    <Button
        android:text="启动BActivity"
        android:layout_width="wrap_content"
        android:layout_height="wrap_content"
        android:id="@+id/button2" />
</LinearLayout>
```

注意，在 activity_b.xml 和 activity_main.xml 两个文件中，3 个控件的设置完全相同。在两个不同文件中，控件 ID 相同并不会造成冲突。

（5）修改 MainActivity.java，代码如下。

```
package com.example.xbg.launchsingletop;
import android.content.Intent;
…
public class MainActivity extends AppCompatActivity {
    @Override
    protected void onCreate(Bundle savedInstanceState) {
        super.onCreate(savedInstanceState);
        setContentView(R.layout.activity_main);
        TextView tv= (TextView) findViewById(R.id.textView);
        tv.setText(String.format("任务ID：%d\n活动实例：%s",getTaskId(),this.toString()));
        findViewById(R.id.button1).setOnClickListener(new View.OnClickListener() {
            @Override
            public void onClick(View v) {
                Intent intent=new Intent(MainActivity.this,MainActivity.class);
                startActivity(intent);
            }
        });
        findViewById(R.id.button2).setOnClickListener(new View.OnClickListener() {
            @Override
            public void onClick(View v) {
                Intent intent=new Intent(MainActivity.this,BActivity.class);
                startActivity(intent);
            }
        });
    }
}
```

（6）修改 BActivity.java，代码如下。

```
package com.example.xbg.launchsingletop;
import android.content.Intent;
…
public class BActivity extends AppCompatActivity {
    @Override
    protected void onCreate(Bundle savedInstanceState) {
        super.onCreate(savedInstanceState);
        setContentView(R.layout.activity_b);
        TextView tv= (TextView) findViewById(R.id.textView);
```

```
        tv.setText(String.format("任务ID：%d\n活动实例：%s",getTaskId(),this.toString()));
        findViewById(R.id.button1).setOnClickListener(new View.OnClickListener() {
            @Override
            public void onClick(View v) {
                Intent intent=new Intent(BActivity.this,MainActivity.class);
                startActivity(intent);
            }
        });
        findViewById(R.id.button2).setOnClickListener(new View.OnClickListener() {
            @Override
            public void onClick(View v) {
                Intent intent=new Intent(BActivity.this,BActivity.class);
                startActivity(intent);
            }
        });
    }
}
```

代码中使用文本视图控件显示任务 ID 和当前活动实例。单击按钮 button1 时启动 MainActivity，单击按钮 button2 时启动 BActivity。MainActivity 和 BActivity 在功能上相同，只是实现代码略有不同，请注意代码中的区别。

（7）运行项目，测试运行效果。

注意，此时并没有设置 MainActivity 和 BActivity 两个活动的启动模式，默认情况下为 standard 模式。所以在运行过程中可发现，不管启动哪一个活动，系统都在同一个任务返回栈中创建新的实例。

（8）停止项目，修改清单文件 AndroidManifest.xml，将 MainActivity 和 BActivity 两个活动的启动模式设置为 singleTop，代码如下。

```
<?xml version="1.0" encoding="utf-8"?>
<manifest xmlns:android="http://schemas.android.com/apk/res/android"
    package="com.example.xbg.launchsingletop">
    <application
        android:allowBackup="true"
        android:icon="@mipmap/ic_launcher"
        android:label="@string/app_name"
        android:supportsRtl="true"
        android:theme="@style/AppTheme">
        <activity android:name=".MainActivity"
            android:launchMode="singleTop">
            <intent-filter>
                <action android:name="android.intent.action.MAIN" />
                <category android:name="android.intent.category.LAUNCHER" />
            </intent-filter>
        </activity>
        <activity android:name=".BActivity" android:label="BActivity"
            android:launchMode="singleTop"></activity>
    </application>
</manifest>
```

（9）重新运行项目，测试运行效果。

项目运行主活动界面如图 2-52 所示。在主活动界面中单击 启动MAINACTIVITY 按钮，会发现不能启动新的 MainActivity，因为活动启动模式为 singleTop。

在主活动界面中单击 启动BACTIVITY 按钮，虽然活动启动模式也为 singleTop，但此时任务返回栈顶部为 MainActivity，所以可启动 BActivity，如图 2-53 所示。从图中可以看出，在 BActivity 活动中，任务 ID 与主活动 ID 相同，只是活动实例不同。

在 BActivity 活动中，因为 BActivity 活动实例已经在任务返回栈顶部运行，单击 启动BACTIVITY 按钮不再创建新的 BActivity 活动实例。但此时可单击 启动MAINACTIVITY 按钮创建新的 MainActivity 活动实例。

也就是说，在当前实例中，因为 MainActivity 和 BActivity 活动的启动模式都为 singleTop，所以不能启动相同活动，只能两个活动交替启动。

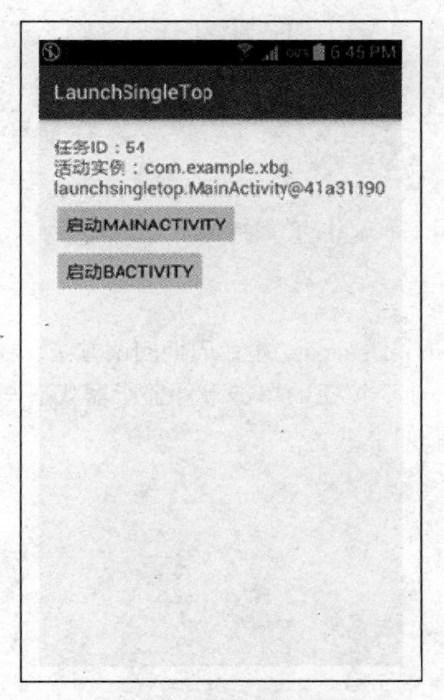

图 2-52　主活动界面

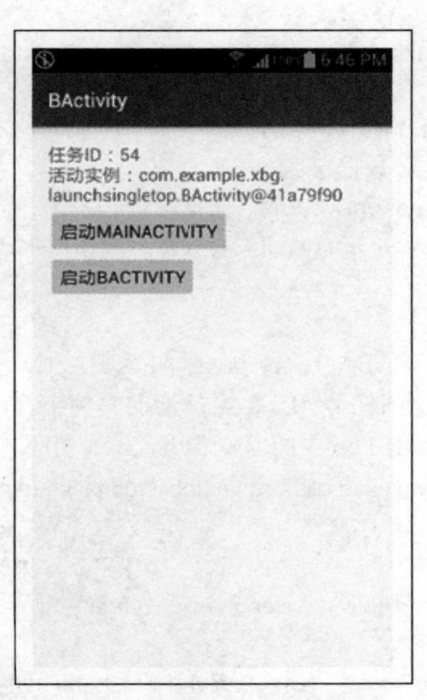

图 2-53　BActivity 活动界面

2.6.3　singleTask 和 singleInstance 模式

singleTask 与 singleInstance 模式

singleTask 启动模式表示一个任务中只能存在活动的一个实例。在启动 singleTask 模式的活动时，系统如果发现任务返回栈中有该活动实例，则将该实例之上的所有活动出栈，使该实例成为栈顶活动；如果任务返回栈中没有活动实例，则创建一个新的活动实例，将其放到栈顶。

singleInstance 启动模式与 singleTop 模式有点类似，但该类型活动只允许“设备”中存在活动的一个实例。在启动 singleInstance 模式的活动时，系统会为活动实例创建一个新的任务返回栈，设备中的所有应用可共享该活动实例。

下面通过实例说明如何使用活动的 singleTask 和 singleInstance 模式，具体操作步骤如下。

（1）创建一个新项目，将应用名称设置为 LaunchSingleTaskAndInstance，并为项目添加一个空活动。

（2）定义主活动布局文件 activity_main.xml 中的控件，方法与实例项目 LaunchSingleTop 中的

activity_main.xml 相同。

（3）为项目添加一个空活动，活动名称为 BActivity，其布局文件中的控件定义与实例项目 LaunchSingleTop 中的 activity_b.xml 相同。

（4）修改 MainActivity.java，在文本视图控件中显示任务 ID 和当前活动实例字符串。两个命令按钮的单击事件监听器代码与实例项目 LaunchSingleTop 中的 MainActivity.java 相同，其中不同的代码如下。

```
package com.example.xbg.launchsingletaskandinstance;
...
public class MainActivity extends AppCompatActivity {
    @Override
    protected void onCreate(Bundle savedInstanceState) {
        ...
        System.out.println( "任务ID："+getTaskId()+"\n"+this.toString()+"正在创建！");
    }
    @Override
    protected void onDestroy() {
        super.onDestroy();
        System.out.println( "任务ID："+getTaskId()+"\n"+this.toString()+"已经销毁！");
    }
}
```

（5）修改 BActivity.java，并为 BActivity.java 添加 onDestroy()生命周期回调方法，在文本视图控件中显示的任务 ID 和当前活动实例字符串以及两个命令按钮的单击事件监听器代码与实例项目 LaunchSingleTop 中的 BActivity.java 相同，不同的代码如下。

```
package com.example.xbg.launchsingletaskandinstance;
...


public class BActivity extends AppCompatActivity {
    @Override
    protected void onCreate(Bundle savedInstanceState) {
        ...
        System.out.println( "任务ID："+getTaskId()+"\n"+this.toString()+"正在创建！");
    }
    @Override
    protected void onDestroy() {
        super.onDestroy();
        System.out.println( "任务ID："+getTaskId()+"\n"+this.toString()+"已经销毁！");
    }
}
```

（6）修改清单文件 AndroidManifest.xml，将 MainActivity 的启动模式设置为 singleTask，将 BActivity 的启动模式设置为 standard，代码如下。

```
<?xml version="1.0" encoding="utf-8"?>
<manifest xmlns:android="http://schemas.android.com/apk/res/android"
    package="com.example.xbg.launchsingletaskandinstance">
    <application
```

```
        android:allowBackup="true"
        android:icon="@mipmap/ic_launcher"
        android:label="@string/app_name"
        android:supportsRtl="true"
        android:theme="@style/AppTheme">
        <activity android:name=".MainActivity"
            android:launchMode="singleTask">
            <intent-filter>
                <action android:name="android.intent.action.MAIN" />
                <category android:name="android.intent.category.LAUNCHER" />
            </intent-filter>
        </activity>
        <activity android:name=".BActivity" android:launchMode="standard"></activity>
    </application>
</manifest>
```

（7）运行项目，测试运行效果。

项目运行时，主活动界面如图 2-54 所示，此时单击 启动MAINACTIVITY 按钮，不会有任何变化；单击 启动BACTIVITY 按钮，可启动 BActivity 活动，界面如图 2-55 所示。

因为 BActivity 活动是 standard 启动模式的，所以可连续多次单击 启动BACTIVITY 按钮，创建多个 BActivity 活动实例。图 2-56 所示为连续 3 次单击 启动BACTIVITY 按钮后，再单击 启动MAINACTIVITY 按钮时，Android Studio 的 Run 窗口中显示的活动生命周期回调方法输出的信息。因为 MainActivity 活动是 singleTask 模式的，任务返回栈中只允许存在一个实例，所以系统在返回主活动界面时，会将任务返回栈中 MainActivity 实例之上的所有活动出栈。此时按返回键会返回设备主屏幕，而不是返回上一个 BActivity 活动。

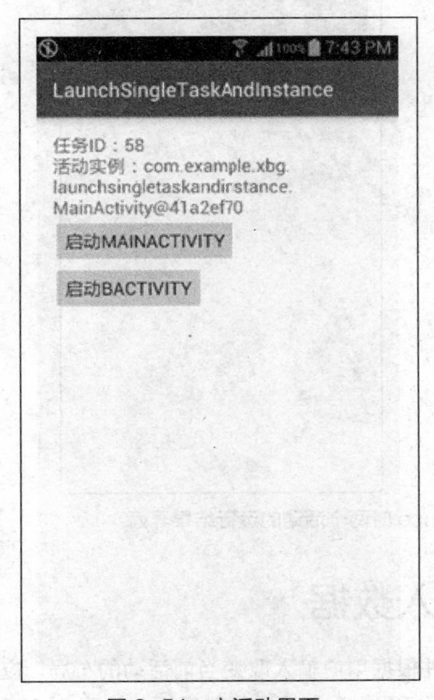

图 2-54 主活动界面

图 2-55 BActivity 活动

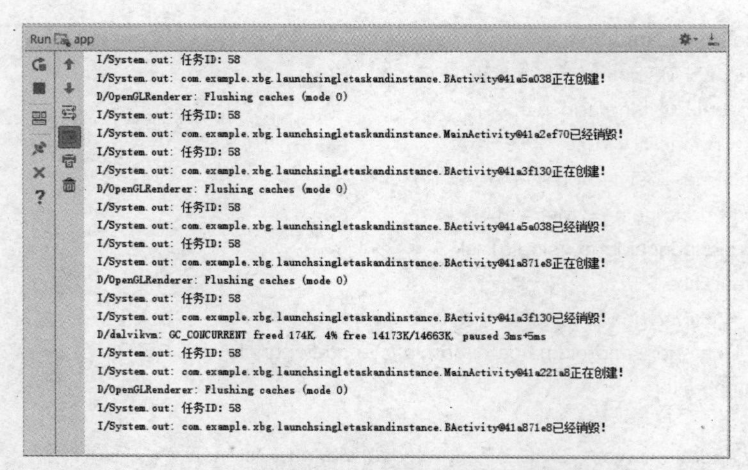

图 2-56　多次启动 BActivity 活动后再启动活动过程中 Run 窗口的信息

（8）修改清单文件 AndroidManifest.xml，将 MainActivity 的启动模式设置为 singleInstance，BActivity 的启动模式仍为 standard，

（9）重新运行项目，测试运行效果。

修改后，在项目运行过程中可发现，MainActivity 活动实例只有一个，而且和 BActivity 活动所在的任务返回栈 ID 不同。多个 BActivity 活动实例的任务返回栈 ID 相同。在测试运行过程中可发现，因为 MainActivity 活动实例和 BActivity 活动实例在不同任务返回栈中，所以在两种活动实例之间的切换时间会比同一个任务返回栈中活动实例之间的切换时间要长。

图 2-57 所示为 MainActivity 活动界面和 BActivity 活动界面，可看到显示的任务 ID 不同。

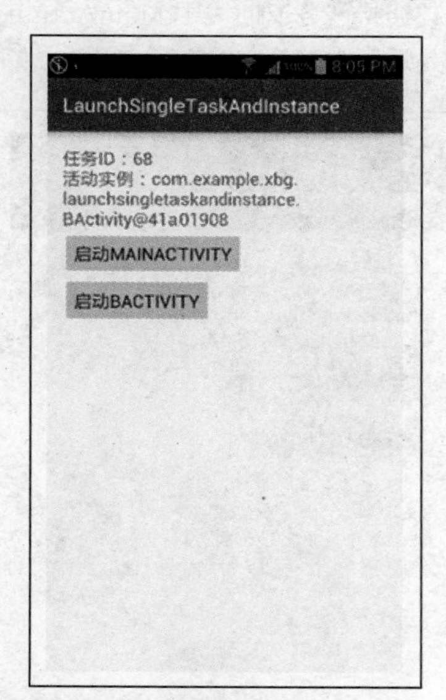

图 2-57　主活动为 singleInstance 时两个活动的运行结果界面

2.7　编程实践：获取用户输入数据

本章将创建一个可接收用户输入数据的应用，并根据用户输入改变当前活动的布局。应用运行时，主活动初始界面如图 2-58 所示；单击 设置 按钮可打开一个布局设置对话框，如图 2-59 所示。

图 2-58　初始界面

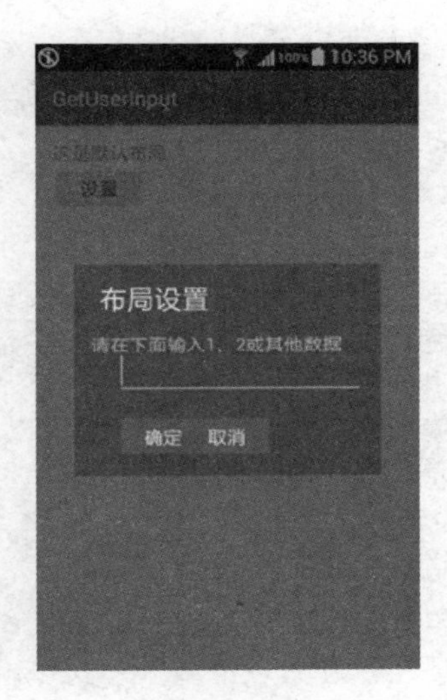

图 2-59　布局设置对话框

在布局设置对话框中可输入 1、2 或其他数据，输入其他数据不改变主活动布局。如果输入的是 1 且主活动不是默认布局，则将其布局设置为默认布局；如果输入的是 2 且主活动不是另一个布局，则将其布局设置为另一个布局。图 2-60 所示分别为默认布局或另一个布局的活动界面。

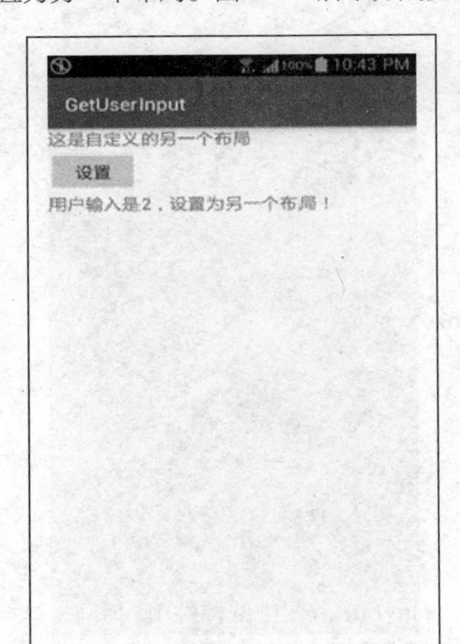

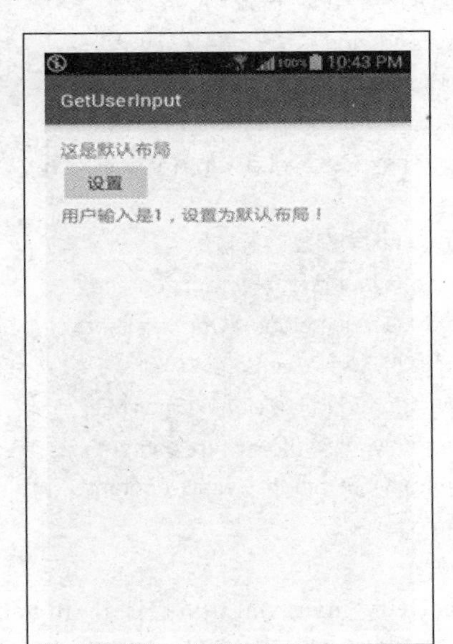

图 2-60　布局发生改变时的活动界面

获取用户输入数据的具体操作步骤如下。

（1）在 Android Studio 中创建一个新项目，将应用名称设置为 GetUserInput，并为项目添加一个空活动。

（2）修改 activity_main.xml，代码如下。

```
<?xml version="1.0" encoding="utf-8"?>
<LinearLayout xmlns:android="http://schemas.android.com/apk/res/android"
    …>
```

```
    <TextView android:id="@+id/textView1"
        android:layout_width="wrap_content"
        android:layout_height="wrap_content"
        android:text="这是默认布局" />
    <Button
        android:text="设置"
        android:layout_width="wrap_content"
        android:layout_height="wrap_content"
        android:id="@+id/btnSet"/>
    <TextView android:id="@+id/textShowMsg"
        android:layout_width="wrap_content"
        android:layout_height="wrap_content"
        android:text="" />
</LinearLayout>
```

（3）为项目添加一个布局资源文件，文件名为 anotherlayout，代码如下。

```
<?xml version="1.0" encoding="utf-8"?>
<LinearLayout xmlns:android="http://schemas.android.com/apk/res/android"
    android:orientation="vertical" android:layout_width="match_parent"
    android:layout_height="match_parent">
    <TextView android:id="@+id/textView1"
        android:layout_width="wrap_content"
        android:layout_height="wrap_content"
        android:text="这是自定义的另一个布局" />
    <Button
        android:text="设置"
        android:layout_width="wrap_content"
        android:layout_height="wrap_content"
        android:id="@+id/btnSet"/>
    <TextView android:id="@+id/textShowMsg"
        android:layout_width="wrap_content"
        android:layout_height="wrap_content"
        android:text="" />
</LinearLayout>
```

注意，activity_main.xml 中的控件 ID 和 anotherlayout.xml 中的控件 ID 相同，这样是为了在活动使用不同布局时使用相同的代码，这充分体现了 Android 中 UI 和代码设计分离的原则。

（4）为项目添加一个空活动，活动名称为 SettingActivity。

（5）修改 activity_setting.xml，为布局添加控件，用于接收用户输入数据，代码如下。

```
<?xml version="1.0" encoding="utf-8"?>
<RelativeLayout xmlns:android="http://schemas.android.com/apk/res/android"
    …>
    <TextView
        android:text="请在下面输入1、2或其他数据"
        android:layout_width="wrap_content"
        android:layout_height="wrap_content"          />
```

```
    <EditText
        android:layout_width="wrap_content"
        android:layout_height="wrap_content"
        android:inputType="textPersonName"
        android:ems="10"
        android:layout_alignParentTop="true"
        android:layout_alignParentLeft="true"
        android:layout_alignParentStart="true"
        android:layout_marginLeft="24dp"
        android:layout_marginStart="24dp"
        android:layout_marginTop="14dp"
        android:id="@+id/editText" />
    <Button
        android:text="确定"
        android:layout_width="wrap_content"
        android:layout_height="wrap_content"
        android:layout_below="@+id/editText"
        android:layout_alignLeft="@+id/editText"
        android:layout_alignStart="@+id/editText"
        android:layout_marginTop="17dp"
        android:id="@+id/btnOk" />
    <Button
        android:text="取消"
        android:layout_width="wrap_content"
        android:layout_height="wrap_content"
        android:layout_alignTop="@+id/btnOk"
        android:layout_centerHorizontal="true"
        android:id="@+id/btnCancel" />
</RelativeLayout>
```

布局中，EditText 用于接收用户输入，按钮 btnOk 用于确认输入，按钮 btnCancel 用于取消输入。只有在用户确认了输入时，才会在主活动中判断是否更改布局。

（6）修改 SettingActivity.java，实现用户确认和取消输入操作，代码如下。

```
package com.example.xbg.getuserinput;
import android.content.Intent;
import android.support.v7.app.AppCompatActivity;
import android.os.Bundle;
import android.view.View;
import android.widget.EditText;
public class SettingActivity extends AppCompatActivity {
    @Override
    protected void onCreate(Bundle savedInstanceState) {
        super.onCreate(savedInstanceState);
        setContentView(R.layout.activity_setting);
        findViewById(R.id.btnOk).setOnClickListener(new View.OnClickListener() {
```

```
            @Override
            public void onClick(View v) {//执行确认操作，返回输入数据
                Intent intent=new Intent();
                EditText editText= (EditText) findViewById(R.id.editText);
                intent.putExtra("userinput",editText.getText().toString());//将输入装入Intent
                setResult(RESULT_OK,intent);//设置返回结果
                finish();//结束当前活动
            }
        });
        findViewById(R.id.btnCancel).setOnClickListener(new View.OnClickListener() {
            @Override
            public void onClick(View v) {//执行取消操作
                Intent intent=new Intent();
                setResult(RESULT_CANCELED,intent);//设置返回结果
                finish();//结束当前活动
            }
        });
    }
}
```

（7）修改 MainActivity.java，添加按钮的单击事件监听器，在单击按钮时，首先创建 Intent 对象，然后在其中封装数据，最后用其启动活动，代码如下。

```
package com.example.xbg.getuserinput;
import android.content.Intent;
import android.support.v7.app.AppCompatActivity;
import android.os.Bundle;
import android.view.View;
import android.widget.EditText;
import android.widget.TextView;
public class MainActivity extends AppCompatActivity {
    private   static int REQUEST_CODE=1000;//设置一个请求码
    private   int currentLayout;//用于记录当前布局ID
    @Override
    protected void onCreate(Bundle savedInstanceState) {
        super.onCreate(savedInstanceState);
        setContentView(R.layout.activity_main);
        currentLayout=R.layout.activity_main;
        initButton();
    }
    private void initButton(){//初始化布局按钮，添加监听器
        findViewById(R.id.btnSet).setOnClickListener(new View.OnClickListener() {
            @Override
            public void onClick(View v) {
                Intent intent=new Intent(MainActivity.this,SettingActivity.class);
                startActivityForResult(intent,REQUEST_CODE);//启动可返回结果的活动
```

```
            }
        });
    }
    @Override
    protected void onActivityResult(int requestCode, int resultCode, Intent data) {
        super.onActivityResult(requestCode, resultCode, data);
        //处理返回结果
        if(requestCode==REQUEST_CODE){//返回的请求码与当前活动请求码一致时，才执行后继操作
            String msg;
            if(resultCode==RESULT_OK){//RESULT_OK表示返回的活动已成功处理请求
                String userinput=data.getStringExtra("userinput");
                msg="用户输入的不是1或2，未改变布局";
                if(userinput.equals("1") & (currentLayout!=R.layout.activity_main)) {
                    setContentView(R.layout.activity_main);
                    currentLayout=R.layout.activity_main;
                    initButton();
                    msg="用户输入的是1，设置为默认布局！";
                }else if(userinput.equals("2")  & (currentLayout==R.layout.activity_main)){
                    setContentView(R.layout.anotherlayout);
                    currentLayout=R.layout.anotherlayout;
                    initButton();
                    msg="用户输入的是2，设置为另一个布局！";
                }else{
                    msg="用户输入的是"+userinput+"，无须改变布局！";
                }
            }
            else{
                msg="用户取消了布局选择操作！";
            }
            TextView tv= (TextView) findViewById(R.id.textShowMsg);
            tv.setText(msg);
        }
    }
}
```

（8）运行项目，测试运行效果。

2.8 小结

本章首先介绍了 Android 系统的核心组件及相关内容——活动（Activity）是什么、如何为活动绑定自定义视图、如何在活动中启动另一个活动和结束活动，然后重点讲解了如何在活动中使用各种类型的 Intent 以及如何利用 Intent 在活动之间传递数据，最后讲解了活动的生命周期以及活动的启动模式等内容。

活动只是 Android 应用开发的一个起点。熟练理解和掌握本章的内容，开发人员可以在应用中更好地使用和管理活动，也可更好地学习后续内容。

2.9　习题

1. 请简述为一个活动绑定自定义视图的基本步骤。
2. 请问在一个活动中启动另一个活动的基本语法格式是什么？
3. Intent 有哪些类型？这些类型之间有何区别？
4. 请简述向启动的活动中传递数据的基本过程。
5. 请问活动在其生命周期内可能有哪些状态？
6. 请问活动在其生命周期内可能会调用哪些生命周期回调方法？
7. 请问活动有哪几种启动模式？

PARTO3

第3章

UI设计

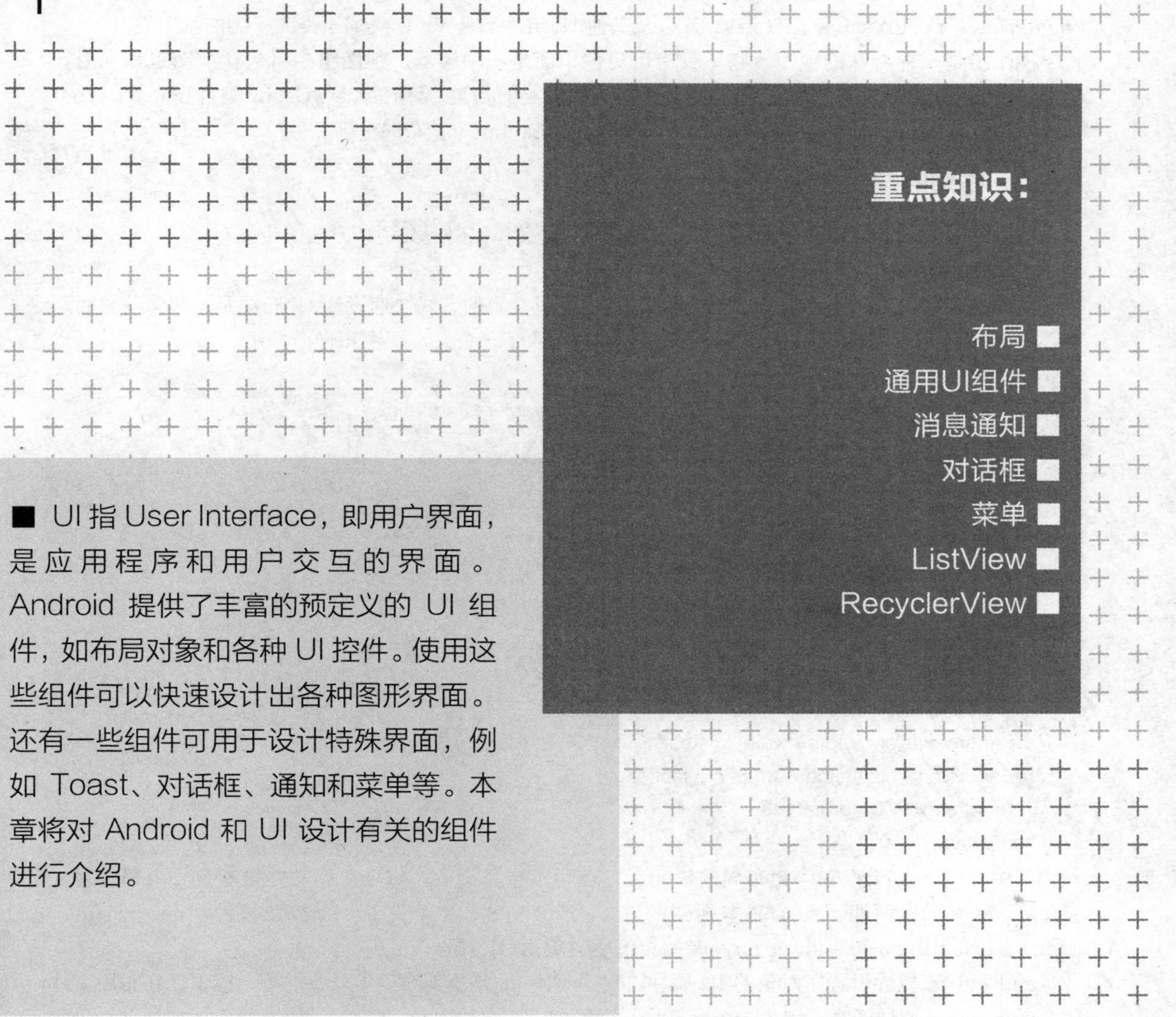

重点知识：

- 布局
- 通用UI组件
- 消息通知
- 对话框
- 菜单
- ListView
- RecyclerView

■ UI指User Interface，即用户界面，是应用程序和用户交互的界面。Android提供了丰富的预定义的UI组件，如布局对象和各种UI控件。使用这些组件可以快速设计出各种图形界面。还有一些组件可用于设计特殊界面，例如Toast、对话框、通知和菜单等。本章将对Android和UI设计有关的组件进行介绍。

3.1 布局

布局是 Android 应用程序的界面定义，布局中的所有界面元素都是视图（View）或视图组（ViewGroup）对象。一个布局首先是一个视图组对象，可在视图组对象中添加子视图组对象或者视图对象。

3.1.1 视图和视图组

视图用于在屏幕上绘制可与用户交互的界面元素。一个视图占据一块矩形屏幕区域，并通过属性设置来渲染此区域。对于视图区域，可设置是否可见、是否可获得焦点，也可处理其中发生的事件（用户触摸、拖动等）。

在 Android 中，View 类是所有用于设计界面组成元素的基类。Button、CheckBox、ExitView、ImageView、ProgressBar、TextView 以及其他 UI 组件都是 View 类的子类或子类的派生类。

视图组是一种特殊的视图，它不具有可见性，而是一种容器。视图组中可包含视图组和视图。ViewGroup 类是 View 类的一个子类，它又是各种布局类的基类。常用的 ViewGroup 类有 LinearLayout（线性布局）、RelativeLayout（相对布局）和 FrameLayout（帧布局）等。

3.1.2 布局的定义方法

在设计 Android 应用程序 UI 时，可通过 XML 定义和代码定义两种方法来定义布局。

1. 布局的 XML 定义

布局的 XML 定义是使用 Android 的 XML 词汇以文本的方式快速设计 UI 布局及其包含的界面元素。例如如下代码，即声明了一个线性布局，布局包含了一个文本视图和一个按钮。

```
<?xml version="1.0" encoding="utf-8"?>
<LinearLayoutxmlns:android="http://schemas.android.com/apk/res/android"
    android:orientation="vertical"
    android:layout_width="match_parent"
    android:layout_height="match_parent">
    <TextView android:id="@+id/textView1"
        android:layout_width="wrap_content"
        android:layout_height="wrap_content"
        android:text="这是自定义的另一个布局" />
    <Button
        android:text="设置"
        android:layout_width="wrap_content"
        android:layout_height="wrap_content"
        android:id="@+id/btnSet"/>
</LinearLayout>
```

其中，<LinearLayout>为布局的根元素，声明了布局类型。每个布局文件只能有一个根元素。<TextView>元素声明了一个文本视图控件，<Button>元素声明了一个按钮控件。android:id、android:orientation 和 android:layout_width 等称为 XML 属性。

android:id 属性设置了对象的 ID 属性。视图对象 ID 是当前布局中对象的唯一标识。在布局文件中，android:id 属性通常设置为一个字符串，示例如下。

```
android:id="@+id/textView1"
```

其中，@符号用于指示 XML 解析程序开始解析 ID 字符串；+（加号）表示添加一个新的 ID；id 表示当前属性为对象的 ID 声明；textView1 为 ID 字符串。

在布局文件中通过 ID 字符串引用对象时，不需要使用加号，示例如下。

```
android:layout_below="@id/textView1"
```

应用程序编译后，对象的 ID 字符串映射为一个整数，在代码中，通过“R.id.ID 字符串”格式来引用对象。例如，如下语句即创建了一个 Button 对象变量来引用布局文件中声明的 Button 对象。

```
Button btnOk=(Button)findViewById(R.id.btnOk);
```

每个视图对象或视图组对象都支持该类的各种 XML 属性，属性决定了布局组件的外观和行为方式。

包含布局定义的 XML 文件称为布局文件，放在应用程序的 res\layout 文件夹中。布局 ID 默认与布局文件名相同。在代码中，通过“R.layout.布局 ID”来引用布局文件。例如，如下语句设置了当前活动绑定的布局。

```
setContentView(R.layout.activity_setting);
```

2. 布局的代码定义

在代码中，可通过创建视图类和视图组类的实例对象来定义布局。Android 推荐使用 XML 布局文件来定义布局。XML 布局文件支持可视化的设计视图，通过拖放操作即可完成布局设计。XML 布局文件也使应用程序的 UI 设计和功能代码分离，体现了现代应用程序设计理念。所以这里不对如何通过代码定义布局进行赘述。

3.1.3 线性布局 LinearLayout

LinearLayout 是一个视图组，它按照垂直或水平方式顺序排列内部的视图或视图组对象。在线性布局中，每行或每列中只允许有一个子视图或视图组对象。

LinearLayout

线性布局的主要 XML 属性如下。

- android:gravity：设置内部组件的显示位置。
- android:orientation：设置内部组件的排列方向，常量 horizontal 表示水平排列，vertical（默认值）表示垂直排列。
- android:background：设置一个 drawable 资源作为背景。
- android:id：设置布局 ID。
- android:padding：设置所有边距的统一值。
- android:paddingBottom：设置底部边距。
- android:paddingLeft：设置左边距。
- android:paddingRight：设置右边距。
- android:paddingTop：设置顶部边距。

宽度、高度、边距等尺寸参数可使用 px（像素）或 dp（独立像素）作为单位。1px 对应屏幕上的一个点。dp 是基于屏幕密度的抽象单位，在每英寸 160 点的屏幕上，1dp=1px。屏幕密度不同，dp 和 px 的换算不同。

在线性布局中，可用 android:layout_weight 属性为各个子视图分配权重。“权重”值更大的视图占用屏幕剩余空间的比例更多。android:layout_weight 属性的默认值为 0，表示视图按实际大小绘制，不进行扩展。

在绘制视图时，布局管理器首先按实际大小绘制权重为 0 的视图，权重不为 0 的视图按照权重计算比例分配屏幕剩余空间。

在线性布局中，如果要让多个视图平均分配屏幕空间，可将每个视图的 android:layout_height（垂直布局）或 android:layout_width（水平布局）设置为 0，然后将每个视图的 android:layout_weight 属性设置为 1 或其他相同的值即可。

如下代码在线性布局中添加了 3 个按钮控件，并设置了不同的权重。实例项目：源代码\03\LinearLayout。

```
<?xml version="1.0" encoding="utf-8"?>
<LinearLayoutxmlns:android="http://schemas.android.com/apk/res/android"
    android:orientation="vertical" android:layout_width="match_parent"
    android:layout_height="match_parent">
    <Buttonandroid:layout_width="match_parent"
        android:layout_height="wrap_content"
        android:text="按钮1" />
    <Buttonandroid:layout_width="wrap_content"
        android:layout_height="wrap_content"
        android:layout_weight="1"
        android:text="按钮2" />
    <Button   android:layout_width="match_parent"
        android:layout_height="wrap_content"
        android:layout_weight="1"
        android:text="按钮3" />
</LinearLayout>
```

此时，第 1 个按钮权重为 0，第 2、3 个按钮权重都为 1，则第 1 个按钮按实际大小绘制，它的宽度为 match_parent，即占满屏幕宽度；它的高度为 wrap_content，即刚好包含内容；第 2、3 个按钮的高度虽然设置为 wrap_content，但权重均为 1，所以会扩展高度，并分别占用剩余屏幕高度的一半，如图 3-1 所示。如果第 3 个按钮权重改为 3，则第 2 个按钮的高度占剩余屏幕空间的 1/4，第 3 个按钮占 3/4，如图 3-2 所示。

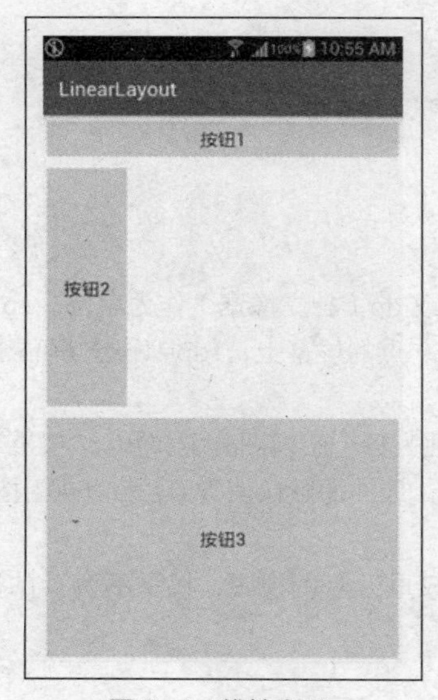

图 3-1　线性布局 1

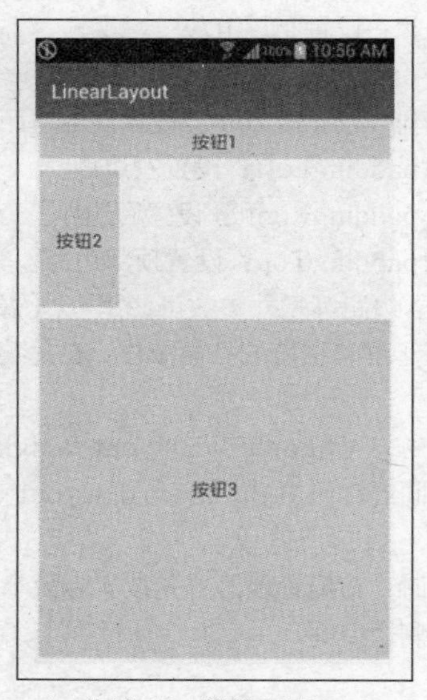

图 3-2　线性布局 2

视图组中也可包含子视图组。例如如下代码，即利用线性布局嵌套设计了一个简单的短信发送界面。实例项目：源代码\03\LinearLayout2。

```
<?xml version="1.0" encoding="utf-8"?>
<LinearLayoutxmlns:android="http://schemas.android.com/apk/res/android"
    xmlns:tools="http://schemas.android.com/tools"
    android:id="@+id/activity_main"     android:layout_width="match_parent"
    android:layout_height="match_parent"     android:orientation="vertical"
    tools:context="com.example.xbg.linearlayout2.MainActivity">
    <LinearLayout android:layout_width="match_parent"
        android:layout_height="wrap_content"
        android:orientation="horizontal">
        <EditText android:id="@+id/txtReceiver"
            android:layout_weight="1"
            android:layout_width="wrap_content"
            android:layout_height="wrap_content"
            android:hint="收件人"/>
        <Button android:id="@+id/btnSave"
            android:layout_width="wrap_content"
            android:layout_height="wrap_content"
            android:text="发送"/>
    </LinearLayout>
    <EditText android:id="@+id/txtContent"
            android:layout_width="match_parent"
            android:layout_height="wrap_content"
            android:background="#E0FFFF"
            android:layout_weight="1"
        android:hint="短信内容"android:gravity="top" />
</LinearLayout>
```

运行以上代码，布局的显示效果如图 3-3 所示。

图 3-3　嵌套视图组实现的简单布局

RelativeLayout

3.1.4　相对布局 RelativeLayout

RelativeLayout 是一个视图组，它按照相对位置来排列各个子视图。

在使用相对布局时，子视图默认位于左上角，可使用下列属性来控制子视图的位置。

- android:layout_alignParentTop：设置为 true 时，子视图的上边框与父视图的上边框对齐。
- android:layout_centerVertical：设置为 true 时，子视图在垂直方向上位于父视图中间位置。
- android:layout_centerHorizontal：设置为 true 时，子视图在水平方向上位于父视图中间位置。
- android:layout_below：设置一个控件 ID，子视图位于该控件下方。
- android:layout_toRightOf：设置一个控件 ID，子视图位于该控件右侧。
- android:layout_toLeftOf：设置一个控件 ID，子视图位于该控件左侧。

RelativeLayout.LayoutParams 类中还定义了其他一些属性，用于设置控件的相对位置，详细内容可查看帮助文档。

如下代码即使用相对布局设计了一个简单的登录界面。实例项目：源代码\03\RelativeLayout。

```
<?xml version="1.0" encoding="utf-8"?>
<RelativeLayoutxmlns:android="http://schemas.android.com/apk/res/android"
    xmlns:tools="http://schemas.android.com/tools" android:id="@+id/activity_main"
    android:layout_width="match_parent"      android:layout_height="match_parent"
    tools:context="com.example.xbg.relativelayout.MainActivity">
    <EditText
        android:layout_width="match_parent"
        android:layout_height="wrap_content"
        android:layout_alignParentTop="true"
        android:hint="输入用户名"
        android:id="@+id/editText1" />
    <EditText
        android:layout_width="match_parent"
        android:layout_height="wrap_content"
        android:layout_below="@id/editText1"
        android:hint="输入密码"
        android:id="@+id/editText2" />
    <Button
        android:layout_width="100dp"
        android:layout_height="wrap_content"
        android:layout_below="@id/editText2"
        android:layout_alignParentRight="true"
        android:text="确定" />
</RelativeLayout>
```

运行以上代码，布局的显示效果如图 3-4 所示。

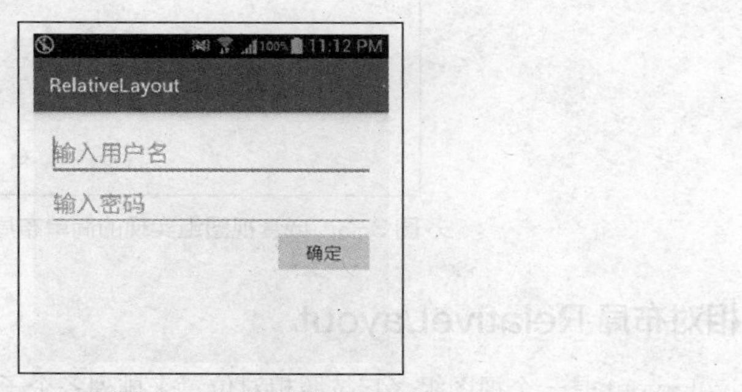

图 3-4　相对布局

3.1.5 帧布局 FrameLayout

FrameLayout

帧布局是一种特殊的布局，它以层叠的方式显示布局中的多个控件，最后添加的控件位于最前面。

默认情况下，控件位于帧布局的左上角，通过控件的 android:layout_gravity 属性可控制其位置。android:layout_gravity 属性可设置为下列值。

- top：控件位于布局顶部。
- bottom：控件位于布局底部，单独使用时等价于“left|bottom”。
- left：控件位于布局左侧。
- right：控件位于布局右侧，单独使用时等价于“top|right”。
- center：控件位于布局中心。
- center_vertical：控件位于垂直方向上的中间位置，单独使用时等价于“left|center_vertical”。
- center_horizontal：控件位于水平方向上的中间位置，单独使用时等价于“top|center_horizontal”。

这些值可以组合使用。默认情况下，控件的 android:layout_gravity 属性具有“left|top”值，即位于左上角。底部居中可表示为“bottom|center”或“bottom|center_horizontal”。

如下代码即在帧布局中添加了两个文本视图和按钮。实例项目：源代码\03\FrameLayout。

```
<?xml version="1.0" encoding="utf-8"?>
<FrameLayoutxmlns:android="http://schemas.android.com/apk/res/android"
    xmlns:tools="http://schemas.android.com/tools"
    android:id="@+id/activity_main"
    android:layout_width="match_parent"      android:layout_height="match_parent"
    tools:context="com.example.xbg.framelayout.MainActivity">
    <TextView
        android:layout_width="wrap_content"
        android:layout_height="wrap_content"
        android:textSize="40dip"
        android:textColor="#ff0000"
        android:text="第3层文本视图"
        android:id="@+id/textView1" />
    <TextView
        android:text="第2层文本视图"
        android:layout_width="wrap_content"
        android:layout_height="wrap_content"
        android:textColor="#000dff"
        android:textSize="30dip"
        android:id="@+id/textView2" />
    <Button
        android:text="第1层按钮"
        android:layout_width="wrap_content"
        android:layout_height="wrap_content"
        android:id="@+id/button" />
</FrameLayout>
```

运行以上代码，布局的显示结果如图 3-5 所示。因为没有设置位置，所以各个控件都显示在屏幕左上角。

为文本视图 textView2 添加如下属性代码，可设置其位于屏幕中心。

```
android:layout_gravity="center"
```

为按钮 button 添加如下属性代码，可设置其位于屏幕底部居中。

```
android:layout_gravity="center|bottom"
```

修改后，布局的运行效果如图 3-6 所示。

图 3-5　默认的帧布局

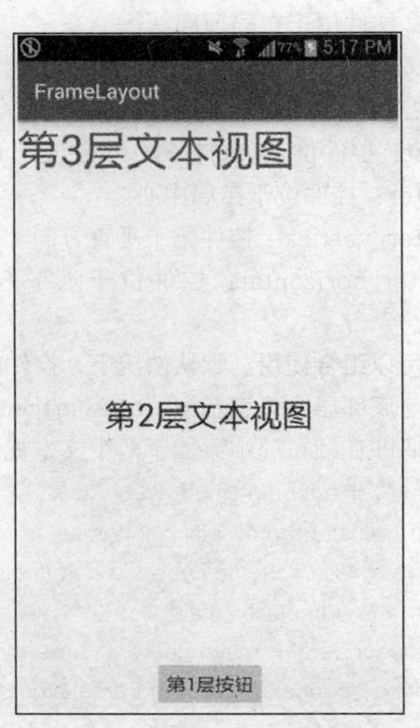

图 3-6　设置位置后的帧布局

3.2　通用 UI 组件

Android 提供了多种可在 UI 中使用的控件，如文本视图（TextView）、按钮（Button、ImageButton）、文本字段（EditText、AutoCompleteTextView）、复选框（CheckBox）、单选按钮（RadioButton）、切换按钮（ToggleButton）、微调框（Spinner）、日期选取器（DatePicker）及时间选取器（TimePicker）等。

3.2.1　文本视图（TextView）

文本视图用于显示指定的文本。例如如下代码，即在布局文件中添加了一个文本视图控件。本节实例项目：源代码\03\LearnUIComponent。

```
<TextView
    android:layout_width="wrap_content"
    android:layout_height="wrap_content"
    android:text="hello，极客学院" />
```

因为没有为文本视图指定 ID，所以它仅仅用于显示指定的文本，不能通过代码访问。

使用如下属性可设置文本显示效果。

- android:typeface：设置字体。Android 默认支持 4 种内置字体，包括 normal、sans、serif 和 monospace。

- android:textSize：设置字号。
- android:textColor：设置颜色。
- android:textStyle：设置文本样式，可设置为 bold、italic 或 bolditalic。

例如如下代码。

```
<TextView
        android:layout_width="wrap_content"
        android:layout_height="wrap_content"
        android:text="hello，极客学院"
        android:typeface="serif"
        android:textSize="20dip"
        android:textColor="#FF0000"/>
```

使用样式可以实现控件的外观设计，达到设计与内容分离的目的，例如如下代码。

```
<TextView style="@style/textViewStyle1" android:text="hello，极客学院"/>
```

其中，textViewStyle1 是在样式文件 styles.xml（位于 res\values 文件夹）中定义的样式，代码如下。

```
<resources>
    <style name="textViewStyle1">
        <item name="android:layout_width">wrap_content</item>
        <item name="android:layout_height">wrap_content</item>
        <item name="android:textColor">#FF0000</item>
        <item name="android:typeface">monospace</item>
        <item name="android:textSize">20dip</item>
    </style>
</resources>
```

3.2.2 按钮（Button、ImageButton）

按钮用于在用户触摸时执行某种操作。Android 允许在按钮中显示文本、图标，或者文本和图标同时显示。

显示文本或文本和图标同时显示时，使用 Button 类来创建按钮，例如如下代码。本节实例项目：源代码\03\LearnUIComponent。

```
<Button
        android:text="Button1"
        android:layout_width="wrap_content"
        android:layout_height="wrap_content"
        android:id="@+id/button1" />
<Button
        android:text="Button2"
        android:layout_width="wrap_content"
        android:layout_height="wrap_content"
        android:drawableLeft="@mipmap/ic_launcher"
        android:id="@+id/button2" />
```

只显示图标时，可使用 ImageButton 类来创建按钮，例如如下代码。

```
<ImageButton
        android:layout_width="wrap_content"
```

```
        android:layout_height="wrap_content"
        android:src="@mipmap/ic_launcher"
        android:id="@+id/imageButton1" />
```

通常需要为按钮添加 Click 事件监听器。一种方法是在 android:onClick 属性中设置 Click 事件监听器，示例如下。

```
<Button
        ...
        android:id="@+id/button1"
        android:onClick="ClickButton1"/>
```

其中的 ClickButton1 是在代码中定义的一个方法，且方法必须是 public 或 void 类型的，示例如下。

```
public void ClickButton1(View view){
        TextView tv1= (TextView) findViewById(R.id.textView);
        tv1.setText("单击按钮Button1");
    }
```

另一种为按钮添加 Click 事件监听器的方法是在代码中执行 setOnClickListener()方法，示例如下。

```
protected void onCreate(Bundle savedInstanceState) {
        super.onCreate(savedInstanceState);
        setContentView(R.layout.activity_main);
        Button bt2=(Button) findViewById(R.id.button2);
        bt2.setOnClickListener(new View.OnClickListener() {
            @Override
            public void onClick(View v) {
                TextView tv1= (TextView) findViewById(R.id.textView);
                tv1.setText("单击按钮Button2");
            }
        });
    }
```

3.2.3 文本字段（EditText、AutoCompleteTextView）

文本字段控件用于接收用户输入，可使用 android:inputType 属性定义各种输入行为准则。常用 android:inputType 属性值如下。

Android 用户界面之常用控件 EditText

- text：允许输入各种文本。
- textMultiLine：允许输入多行文本。
- textEmailAddress：只允许输入 E-mail 地址。
- textPassword：用于输入密码。
- number：只允许输入数字。
- phone：用于输入电话号码。
- datetime：用于输入日期时间。

例如，如下代码即为布局添加了一个用于输入密码的文本字段控件。本节实例项目：源代码\03\LearnUIComponent。

```
<EditText
        android:layout_width="wrap_content"
        android:layout_height="wrap_content"
        android:inputType="textPassword"
```

```
    android:id="@+id/editText" />
```

AutoCompleteTextView 用于创建提供自动完成提示功能的文本字段控件。创建提供自动完成功能的文本字段控件通常有以下几个步骤。

第 1 步：在布局文件中添加 AutoCompleteTextView 控件，示例如下。

```
<AutoCompleteTextView
        android:layout_width="match_parent"
        android:layout_height="wrap_content"
        android:completionThreshold="1"
        android:id="@+id/autoCompleteTextView"    />
```

属性 android:completionThreshold 设置输入几个字符时显示自动完成提示。

第 2 步：在资源文件 res/values/strings.xml 中定义提供自动完成提示功能的字符串数组资源，示例如下。

```
<resources>
...
    <string-array name="select_array">
        <item>cable</item>
        <item>china</item>
        <item>Chinese</item>
        <item>Check</item>
        </string-array>
        </resources>
```

第 3 步：为 AutoCompleteTextView 绑定提供自动完成提示功能的适配器。例如：

```
AutoCompleteTextView act=(AutoCompleteTextView)findViewById(R.id.autoCompleteTextView);
String[] selects = getResources().getStringArray(R.array.select_array);
ArrayAdapter<String> adapter =
                        newArrayAdapter<String>(this, android.R.layout.simple_list_item_1, selects);
act.setAdapter(adapter);
```

运行示例代码，即可显示自动完成提示功能的文本字段控件，如图 3-7 所示。

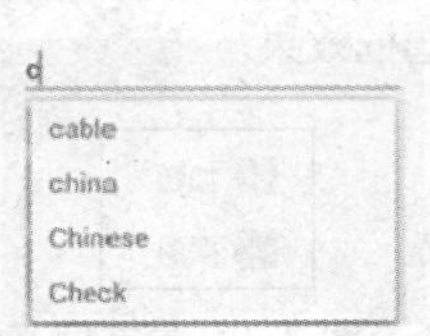

图 3-7　显示了自动完成提示的文本字段控件

3.2.4　复选框（CheckBox）

复选框用于显示一组选项，并允许用户同时选中一个或多个选项。例如，如下代码即为布局添加两个复选框。本节实例项目：源代码\03\LearnUIComponent。

Android 用户界面之常用控件 CheckBox

```
<CheckBox
        android:text="加粗"
        android:layout_width="match_parent"
        android:layout_height="wrap_content"
        android:id="@+id/checkBox1" android:onClick="ClickCheckBox1" />
```

```
    <CheckBox
        android:text="倾斜"
        android:layout_width="match_parent"
        android:layout_height="wrap_content"
        android:id="@+id/checkBox2" android:onClick="ClickCheckBox2" />
```

android:onClick 属性为复选框绑定了 Click 事件监听器，处理复选框 Click 事件。例如，如下代码可实现在单击复选框时改变文本视图的样式。

```
private boolean checked1;
public void ClickCheckBox1(View view){
        checked1 = ((CheckBox) view).isChecked();
        ChangeTextViewStyle();
    }
    private boolean checked2;
    public void ClickCheckBox2(View view){
        checked2 = ((CheckBox) view).isChecked();
        ChangeTextViewStyle();
    }
    public void ChangeTextViewStyle(){
        TextView tv1= (TextView) findViewById(R.id.textView);
        Typeface tf= tv1.getTypeface();
        int style=0;
        if(checked1){
            style=1;
            if(checked2){style=3;}
        }
        else
            if(checked2){style=2;}
        tv1.setTypeface(tf,style);
}
```

图 3-8 所示为上述程序运行时的复选框效果。

图 3-8　复选框

Android 用户界面之常用控件 RadioButton

3.2.5　单选按钮（RadioButton）

单选按钮用于创建一组选项，一次只能选中其中的一项。RadioGroup 作为单选按钮容器，其中的所有单选按钮为一个组。

例如，如下代码在布局中添加一组颜色选项。本节实例项目：源代码\03\LearnUIComponent。

```
<RadioGroup
        android:layout_width="match_parent"
        android:layout_height="match_parent"
        android:orientation="horizontal"
```

```
        android:checkedButton="@+id/radioButton1" >
        <RadioButton
            android:text="蓝色"
            android:layout_width="wrap_content"
            android:layout_height="wrap_content"
            android:id="@+id/radioButton1"
            android:layout_weight="1" android:onClick="ClickRadio" />
        <RadioButton
            android:text="红色"
            android:layout_width="wrap_content"
            android:layout_height="wrap_content"
            android:id="@+id/radioButton2"
            android:layout_weight="1" android:onClick="ClickRadio"/>
        <RadioButton
            android:text="绿色"
            android:layout_width="wrap_content"
            android:layout_height="wrap_content"
            android:id="@+id/radioButton3"
            android:layout_weight="1" android:onClick="ClickRadio"/>
</RadioGroup>
```

其中，android:orientation 属性指定单选按钮组的排列方向；android:checkedButton 属性设置选项组中默认选中的单选按钮；android:onClick 属性为单选按钮绑定 Click 事件监听器。例如，如下代码可更改 TextView 控件的颜色。

```
public void    ClickRadio(View view){
        TextView tv1= (TextView) findViewById(R.id.textView);
        switch(view.getId()) {
            case R.id.radioButton1:
                tv1.setTextColor(Color.rgb(0,0,255));
                break;
            case R.id.radioButton2:
                tv1.setTextColor(Color.rgb(255,0,0));
                break;
            case R.id.radioButton3:
                tv1.setTextColor(Color.rgb(0,255,0));
        }
    }
```

图 3-9 所示为程序运行时的单选按钮效果。

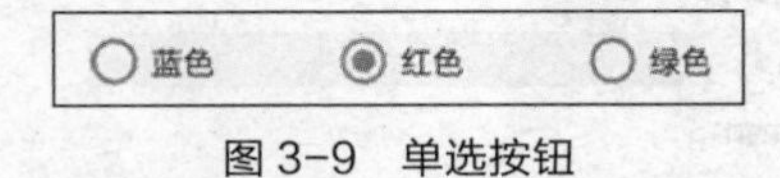

图 3-9　单选按钮

3.2.6　切换按钮（ToggleButton）

切换按钮可以创建一个具有两种状态的按钮。默认情况下，未选中时显示 Off(中文环境为“关闭”)，选中时显示 On（中文环境为“打开”）。android:textOff 属性可设置 Off 状态时显示的文本，

android:textOn 属性可设置 On 状态时显示的文本。

例如，如下代码为布局添加一个切换按钮。本节实例项目：源代码\03\LearnUIComponent。

```
<ToggleButton
        android:textOff="显示背景图片"android:textOn="隐藏背景图片"
        android:layout_width="wrap_content"android:layout_height="wrap_content"
        android:id="@+id/toggleButton"   />
```

在代码中可调用 setOnCheckedChangeListener()方法为切换按钮绑定事件监听器，处理其状态变化。如下代码利用切换按钮设置或清除布局背景。

```
ToggleButton toggle = (ToggleButton) findViewById(R.id.toggleButton);
        toggle.setOnCheckedChangeListener(new CompoundButton.OnCheckedChangeListener() {
            public void onCheckedChanged(CompoundButtonbuttonView, boolean isChecked) {
                LinearLayout layout=(LinearLayout)findViewById(R.id.activity_main);
                if (isChecked) {
                    layout.setBackgroundResource(R.drawable.back); //为布局设置背景图片
                } else {
                    layout.setBackgroundResource(0);//清除布局背景
                }
            }
        });
```

Android 用户界面之常用控件 Spinner

3.2.7 微调框（Spinner）

微调框也称为下拉列表，它提供一组预定选项供用户选择。默认情况下，微调框只显示当前选中项。微调框与 AutoCompleteTextView 类似，但微调框不提供输入功能，只在被触摸时展开。

如下代码可以为布局添加一个微调框。本节实例项目：源代码\03\LearnUIComponent2。

```
<Spinner
        android:layout_width="match_parent"
        android:layout_height="wrap_content"
        android:id="@+id/spinner" />
```

微调框的选项列表需要使用适配器来创建。如果使用字符串数组资源，则可用 ArrayAdapter；如果使用数据库查询结果，则可用 CursorAdapter。

例如，如下代码在字符串资源文件中定义字符串数组资源。

```
<resources>
...
    <string-array name="spinner_array">
        <item>Java软件开发</item>
        <item>C++软件开发</item>
        <item>Android游戏设计</item>
        <item>UI设计</item>
    </string-array>
</resources>
```

然后在 Activity 中通过代码为微调框创建适配器，代码如下。

```
Spinner spinner = (Spinner) findViewById(R.id.spinner);
```

```
ArrayAdapter<CharSequence>spadapter = ArrayAdapter.createFromResource(this,
                    R.array.spinner_array, android.R.layout.simple_spinner_item);
spadapter.setDropDownViewResource(android.R.layout.simple_spinner_dropdown_item);
spinner.setAdapter(spadapter);
```

另外，也可使用 android:entries 属性设置，例如如下代码。

```
<Spinner
        android:layout_width="match_parent"
        android:layout_height="wrap_content"
        android:entries="@array/spinner_array"
        android:id="@+id/spinner" />
```

当用户从微调框下拉列表中选择一个选项时，微调框会收到一个 on-item-selected 事件。要使微调框处理选择事件，可使用 AdapterView.OnItemSelectedListener 接口以及 onItemSelected()回调方法，例如如下代码。

```
public class MainActivity extends AppCompatActivity implements AdapterView.OnItemSelectedListener {
...
    @Override
    protected void onCreate(Bundle savedInstanceState) {
...
        Spinner spinner = (Spinner) findViewById(R.id.spinner);
        spinner.setOnItemSelectedListener(this);  //绑定微调框选择事件处理程序
    }
...
   @Override
   public void onItemSelected(AdapterView<?> parent, View view, int position, long id) {
       //在微调框下拉列表中选择一项时，将其显示到文本视图中
       TextView tv1= (TextView) findViewById(R.id.textView);
       tv1.setText(parent.getSelectedItem().toString());
    }
    @Override
    public void onNothingSelected(AdapterView<?> parent) {
       //
    }
}
```

图 3-10 所示为程序运行时的微调框及展开的列表效果。

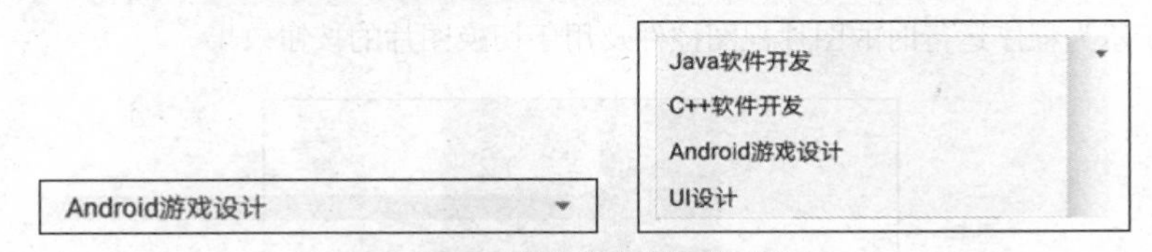

图 3-10　微调框及展开的列表

如果是通过 Activity 实现 AdapterView.OnItemSelectedListener 接口，则可在 Activity 中使用如下语句为微调框绑定选择事件处理程序。

```
spinner.setOnItemSelectedListener(this);
```

如果不实现 OnItemSelectedListener 接口，则可直接创建 OnItemSelectedListener，然后调用

setOnItemSelectedListener()方法来绑定到微调框，例如如下代码。

```
spinner.setOnItemSelectedListener(new AdapterView.OnItemSelectedListener() {
        @Override
        public void onItemSelected(AdapterView<?> parent, View view, int position, long id) {
            TextView tv1= (TextView) findViewById(R.id.textView);
            tv1.setText(parent.getSelectedItem().toString());
        }
        @Override
        public void onNothingSelected(AdapterView<?> parent) {

        }
    });
```

3.2.8 图片视图（ImageView）

图片视图控件用于在界面中显示图片。使用图片视图控件时，需准备图片，并将图片放在资源文件夹 drawable 中。

例如，如下代码即可为布局添加一个图片视图控件。本节实例项目：源代码\03\LearnUIComponent2。

```
<ImageView
    android:layout_width="match_parent"
    android:layout_height="wrap_content"
    android:src="@drawable/run"
    android:id="@+id/imageView" />
```

图片视图控件的 android:src 属性为控件指定显示的图片，也可通过在代码中调用图片视图控件的 setImageResource()方法设置控件显示的图片，例如如下代码。

```
public void changePic(View view){
    ImageView im= (ImageView) findViewById(R.id.imageView);
    imgno++;
    if(imgno%2==0){
        im.setImageResource(R.drawable.run);
    }else{
        im.setImageResource(R.drawable.munt);
    }
}
```

图 3-11 所示为程序运行时的图片视图控件及用于切换图片的按钮效果。

图 3-11　图片视图控件及用于切换图片的按钮

3.2.9 进度条（ProgressBar）

进度条控件通常用于表示程序正在后台处理数据，避免用户枯燥地等待。例如，如下代码即可为布局添加一个进度条。本节实例项目：源代码\03\LearnUIComponent2。

Android 用户界面之常用控件 ProgressBar

```
<ProgressBar
        android:layout_width="match_parent"
        android:layout_height="wrap_content"
        android:id="@+id/progressBar" />
```

这里没有为进度条指定样式，使用默认大小的圆形旋转样式，如图 3-12 所示。

图 3-12　默认样式的进度条

进度条有大图标（progressBarStyleLarge）、中等图标（默认样式，progressBarStyle）、小图标（progressBarStyleSmall）和水平条（progressBarStyleHorizontal）4 种样式，用 style 属性可设置进度条样式，例如如下代码。

```
<ProgressBar
        style="?android:attr/progressBarStyleLarge"
        android:layout_width="match_parent"
        android:layout_height="wrap_content"
        android:id="@+id/progressBar2" />
```

大图标、小图标和水平条样式的进度条如图 3-13 所示。

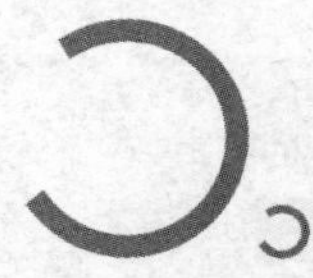

图 3-13　大图标、小图标和水平条样式的进度条

3.2.10 拖动条（SeekBar）

拖动拖动条滑块可获得标识的数值。如下代码即可为布局添加一个拖动条，并将拖动条的最大值设置为 100。本节实例项目：源代码\03\LearnUI Component2。

Android 用户界面之常用控件 SeekBar

```
<SeekBar
        android:layout_width="match_parent"
        android:layout_height="wrap_content"
        android:max="100"
        android:id="@+id/seekBar" />
```

在代码中，可调用 setOnSeekBarChangeListener()方法为拖动条控件绑定拖动条滑块位置变化事件处理程序，例如如下代码。

```
SeekBarsb=(SeekBar)findViewById(R.id.seekBar);
        sb.setOnSeekBarChangeListener(new SeekBar.OnSeekBarChangeListener() {
            @Override
            public void onProgressChanged(SeekBarseekBar, int progress, boolean fromUser) {
```

```
                //拖动滑块时调用
                TextView tv= (TextView) findViewById(R.id.textView2);
                tv.setText("当前拖动条值："+progress);
            }
            @Override
            public void onStartTrackingTouch(SeekBarseekBar) {
                //开始拖动滑块时调用
            }
            @Override
            public void onStopTrackingTouch(SeekBarseekBar) {
                //结束拖动滑块时调用
            }
        });
```

图 3-14 所示为上述程序运行时的拖动条效果。

当前拖动条值：46

图 3-14 拖动条

3.3 消息通知

弹出通知 Toast

在 Android 应用中，可采用 Toast 或 Notification 两种方式向用户提供消息通知。

3.3.1 使用 Toast

Toast 是在应用运行期间，通过类似于对话框的方式向用户显示消息提示。Toast 只占用很少的屏幕，并会在一段时间后自动消失。

例如，如下代码即可创建并显示 Toast 通知。

```
Context context=getApplicationContext();                    //获得应用上下文
String text="这是一个较长时间的Toast";                       //准备Taost中显示的文本
int duration=Toast.LENGTH_LONG;                             //用于设置Toast显示时间
Toast toast=Toast.makeText(context,text,duration);          //生成Toast对象
toast.show();                                               //显示Toast通知
```

代码通过 Toast.makeText(context, text, duration)来创建 Toast 对话框，参数 context 为应用运行环境的上下文，在主活动中也可直接使用 MainActivity.this 来获得应用上下文；参数 text 是 Toast 中显示的文本；参数 duration 设置 Toast 通知显示的时间。常量 Toast.LENGTH_LONG 表示显示较长时间，5s 左右。常量 Toast.LENGTH_SHORT 表示显示较短时间，3s 左右。toast.show()方法用于显示 Toast 通知。

使用 Toast 类的构造函数可生成 Toast 对象，例如如下代码。

```
Toast toast= new Toast(getApplicationContext());
```

用如上方法创建 Toast 对象后，可调用相应的方法来设置显示文本和时间，例如如下代码。

```
toast.setText("这是一个较长时间的Toast");
toast.setDuration(Toast.LENGTH_SHORT);
```

除了显示简单的 Toast 通知外，也可用于显示自定义的视图。下述代码即为首先定义的布局文件 toastlayout。

```
<?xml version="1.0" encoding="utf-8"?>
<LinearLayoutxmlns:android="http://schemas.android.com/apk/res/android"
    xmlns:app="http://schemas.android.com/apk/res-auto"
```

```
    android:orientation="vertical" android:layout_width="match_parent"
    android:layout_height="match_parent"
    android:id="@+id/toast_layout">
    <ImageView
        android:layout_width="match_parent"
        android:layout_height="wrap_content"
        app:srcCompat="@drawable/ic_launcher"
        android:id="@+id/imageView" />
    <TextView
        android:text="自定义视图"
        android:layout_width="match_parent"
        android:layout_height="wrap_content"
        android:id="@+id/textView"
        android:gravity="center" />
</LinearLayout>
```

然后，用自定义的布局文件作为 Toast 对象视图。

```
Toast toast= new Toast(getApplicationContext());
toast.setDuration(Toast.LENGTH_SHORT);
LayoutInflater inflater = getLayoutInflater();
View layout = inflater.inflate(R.layout.toastlayout,
        (ViewGroup) findViewById(R.id.toast_layout));             //将布局文件转换为视图对象
TextViewtextView= (TextView) layout.findViewById(R.id.textView);
textView.setText("这是一个自定义视图的Toast");                    //设置布局中文本视图控件显示的文本
toast.setView(layout);                                            //设置Toast显示的视图
```

LayoutInflater 类的 inflate()方法可将 XML 布局文件转换为一个 View 对象，转换后可调用 findViewById()方法来引用布局中的控件，为控件设置相关属性。通过这种方式，可在代码中有效控制 Toast 中显示的内容。

默认情况下，Toast 显示在屏幕底部居中的位置，调用 setGravity()方法可设置 Toast 的显示位置，例如如下代码：

```
toast setGravity(Gravity.CENTER_VERTICAL, 0, 0);          //设置Toast在屏幕中心
```

代码中，常量 Gravity.CENTER_VERTICAL 表示在屏幕上垂直居中。Gravity 类还提供了其他位置常量，例如 Gravity.TOP、Gravity.LEFT 等。setGravity()方法的第 2 和 3 个参数分别表示在 x 轴和 y 轴的偏移量。

如下代码说明了如何在单击按钮时显示 Toast 通知。本节实例项目：源代码\03\LearnToast。

```
package com.example.xbg.learntoast;
import android.Manifest;
...
public class MainActivity extends AppCompatActivity {
    @Override
    protected void onCreate(Bundle savedInstanceState) {
        super.onCreate(savedInstanceState);
        setContentView(R.layout.activity_main);
        findViewById(R.id.button1).setOnClickListener(new View.OnClickListener() {
            @Override
            public void onClick(View v) {
                Toast toast=Toast.makeText(MainActivity.this,
```

```
                    "这是一个较短时间的Toast",Toast.LENGTH_SHORT);
            toast.show();                                          //显示Toast
        }
    });
    findViewById(R.id.button2).setOnClickListener(new View.OnClickListener() {
        @Override
        public void onClick(View v) {
            Toast toast=Toast.makeText(MainActivity.this,
                    "这是一个较长时间的Toast",Toast.LENGTH_SHORT);
            toast.show();                                          //显示Toast
        }
    });
    findViewById(R.id.button3).setOnClickListener(new View.OnClickListener() {
        @Override
        public void onClick(View v) {
            Toast toast= new Toast(MainActivity.this);             //生成Toast对象
            LayoutInflater inflater = getLayoutInflater();
            View layout = inflater.inflate(R.layout.toastlayout,
                    (ViewGroup) findViewById(R.id.toast_layout)); //将布局文件转换为视图对象
            TextViewtextView= (TextView) layout.findViewById(R.id.textView);
            textView.setText("这是一个自定义视图的Toast");//设置布局中文本视图控件显示的文本
            toast.setView(layout);                                 //设置Toast显示的视图
            toast.setDuration(Toast.LENGTH_LONG);                  //设置视图显示时间
            toast.show();                                          //显示Toast
        }
    });
  }
}
```

上述程序运行时，主活动界面如图 3-15 所示。单击按钮“显示较短时间的 Toast”，效果如图 3-16 所示；单击按钮“显示较长时间 Toast”，效果如图 3-17 所示；单击按钮“显示自定义视图的 Toast”，效果如图 3-18 所示。

图 3-15　主活动界面

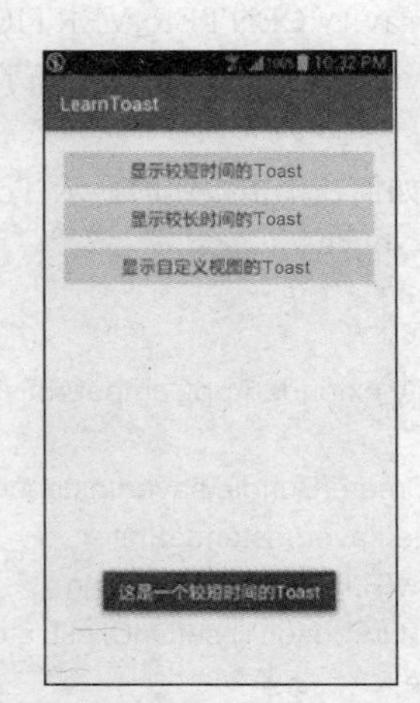

图 3-16　显示较短时间的 Toast

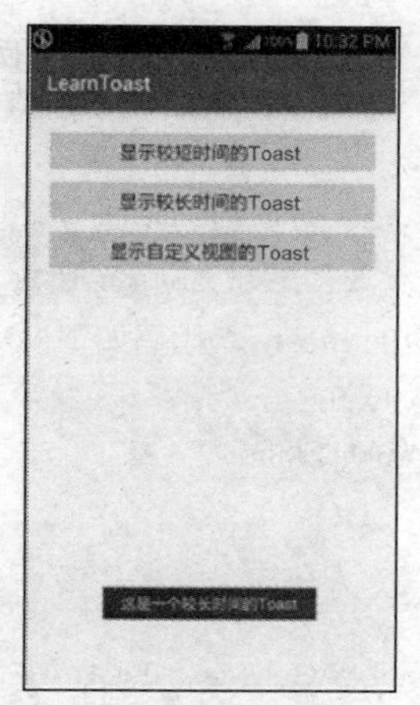

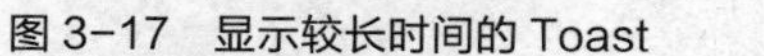
图 3-17　显示较长时间的 Toast　　图 3-18　显示带自定义视图的 Toast

3.3.2　使用 Notification

状态栏提示 Notification

Notification 通知首先在通知区域（也称状态栏）中显示通知图标，展开抽屉式通知栏，可查看通知的详细信息。一个通知通常由图标、标题和内容等组成，图 3-19 所示为一个通知区域，图 3-20 所示为抽屉式通知栏。

图 3-19　通知区域

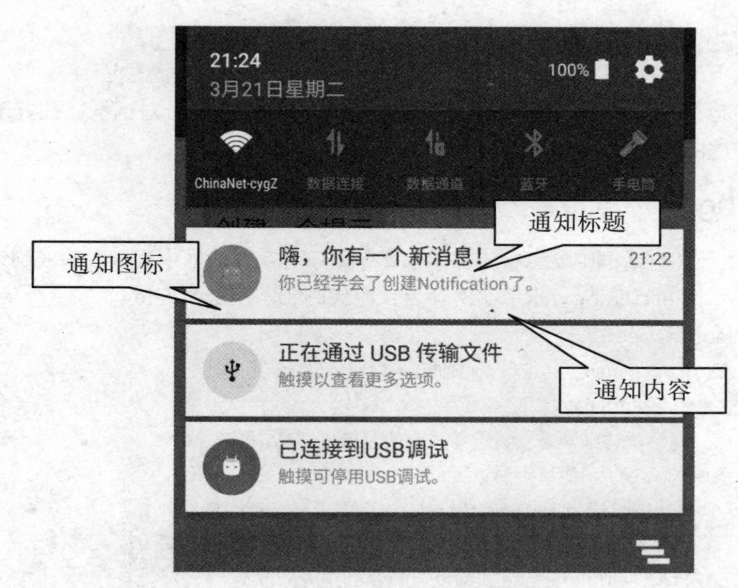

图 3-20　抽屉式通知栏

创建一个简单的通知通常包含如下步骤。

第一步：创建 NotificationCompat.Builder 对象，例如如下代码。本节实例项目：源代码\03\LearnNotification。

```
NotificationCompat.Builder builder=new NotificationCompat.Builder(MainActivity.this);
```

第二步：调用 NotificationCompat.Builder 对象方法设置通知的相关内容，例如如下代码。

```
builder.setSmallIcon(R.drawable.smallico);                    //设置通知小图标
builder.setContentTitle("嗨，你有一个新消息！");              //设置通知标题
builder.setContentText("你已经学会了创建Notification了。");   //设置通知内容
builder.setAutoCancel(true);                                  //设置自动删除通知
```

第三步：创建在抽屉式通知栏中单击通知时启动活动的 Intent，代码如下。

```
Intent resultIntent=new Intent(MainActivity.this,NotificationActivity.class);
TaskStackBuilderstackBuilder=TaskStackBuilder.create(MainActivity.this);
stackBuilder.addParentStack(NotificationActivity.class);
stackBuilder.addNextIntent(resultIntent);
PendingIntentresultPendingIntent=
            stackBuilder.getPendingIntent(0,PendingIntent.FLAG_UPDATE_CURRENT);
builder.setContentIntent(resultPendingIntent);
```

其中，MainActivity 是当前活动类的名称，NotificationActivity 是单击通知时启动的活动类名称。如果仅仅需要显示通知内容，不提供通知单击响应，则该步骤可以省略。

第四步：创建 Notification 对象，代码如下。

```
Notification notification=builder.build();
```

第五步：创建 NotificationManager 对象显示通知，代码如下。

```
NotificationManager manager=
            (NotificationManager) getSystemService(Context.NOTIFICATION_SERVICE);
manager.notify(NOTIFICATION_ID,notification);
```

3.4 对话框

对话框用于在程序运行期间显示一些重要信息，或者用于与用户交互。对话框往往置于界面最前面，会屏蔽其他所有控件的交互能力。

3.4.1 AlertDialog

AlertDialog 对象用于在对话框中显示警告提示信息，用户可在对话框中选择取消或确认操作。例如，如下代码即用于显示 AlertDialog。本节实例项目：源代码\03\LearnDialog。

```
public void showAlertDialog(View view){
        AlertDialog.Builder dialog=new AlertDialog.Builder(MainActivity.this);
        dialog.setTitle("这是一个AlertDialog");
        dialog.setMessage("对话框详细消息：请选择“取消”还是“确认”？");
        dialog.setCancelable(false);//不能取消对话框
        dialog.setPositiveButton("确认", new DialogInterface.OnClickListener() {
            @Override
            public void onClick(DialogInterface dialog, int which) {
                TextView tv1= (TextView) findViewById(R.id.textView);
                tv1.setText("你选择了“确认”！");
            }
        });
        dialog.setNegativeButton("取消", new DialogInterface.OnClickListener() {
            @Override
```

```
            public void onClick(DialogInterface dialog, int which) {
                TextView tv1= (TextView) findViewById(R.id.textView);
                tv1.setText("你选择了 “取消”！ ");
            }
        });
        dialog.show();
    }
```

AlertDialog.Builder 创建了一个 AlertDialog 对象，然后调用相应的方法设置标题、消息、是否可取消（可取消指按返回键或单击对话框之外的空白可取消对话框）等属性。setPositiveButton()方法可设置对话框确认按钮显示的文字，并绑定单击事件处理程序。setNegativeButton()方法可设置对话框取消按钮显示的文字，并绑定单击事件处理程序。

上述程序运行时，显示的对话框如图 3-21 所示。

图 3-21　AlterDialog

3.4.2　ProgressDialog

ProgressDialog 与 AlertDialog 类似，都可弹出一个对话框，并具有屏蔽其他控件的交互能力。ProgressDialog 可在对话框中显示一个进度条。

例如，如下代码即用于显示 ProgressDialog。本节实例项目：源代码\03\LearnDialog。

Android 用户界面之常用控件 ProgressDialog

```
public void    showProgressDialog(View view){
        ProgressDialogprogressDialog=new ProgressDialog(MainActivity.this);
//创建对话框
        progressDialog.setTitle("这是一个进度条对话框");//设置标题
        progressDialog.setMessage("请耐心等待，正在处理数据……");//设置消息
        progressDialog.setCancelable(true);//设置可取消
        progressDialog.show();//显示对话框
    }
```

程序运行时显示的进度条对话框如图 3-22 所示。

这是一个进度条对话框

请耐心等待，正在处理数据......

图 3-22　进度条对话框

日期选取器

3.4.3 DatePickerDialog

DatePickerDialog 用于显示日期选取对话框，定义 OnDateSetListener 监听器，使用 onDateSet()方法可获得在对话框中选取的日期。

例如，如下代码即使用 DatePickerDialog 显示日期选取对话框，并将选取的日期显示在 TextView 中。本节实例项目：源代码\03\LearnDialog。

```
public void    showDateDialog(View view){
        //定义监听器
        DatePickerDialog.OnDateSetListeneronDateSetListener=
                                        newDatePickerDialog.OnDateSetListener() {
            @Override
            public void onDateSet(DatePicker view, int year, int month, int dayOfMonth) {
                TextView tv1= (TextView) findViewById(R.id.textView);
                String theDate=String.format("你选择的日期：%d年%d月#d日",
                                                        year,month,dayOfMonth);
                tv1.setText(theDate);
            }
        };
        //定义DatePickerDialog对象
        DatePickerDialog datePickerDialog=
                        newDatePickerDialog(MainActivity.this,onDateSetListener,2017,5,1);
        datePickerDialog.show();//显示日期选取对话框
    }
```

上述代码中，DatePickerDialog()构造函数的参数依次为当前活动上下文、OnDateSetListener 监听器、初始年、初始月、初始日，程序运行时显示的日期选取对话框如图 3-23 所示。

图 3-23　日期选取对话框

时间选取器

3.4.4 TimePickerDialog

TimePickerDialog 用于显示时间选取对话框，定义 OnTimeSetListener 监听器，使用 onTimeSet()方法可获得在对话框中选取的时间。

例如，如下代码即使用 TimePickerDialog 显示时间选取对话框，并将选取的时间显示在 TextView 中。本节实例项目：源代码\03\LearnDialog。

```
public void    showTimeDialog(View view){
        //定义监听器
        TimePickerDialog.OnTimeSetListeneronTimeSetListener=
                                                newTimePickerDialog.OnTimeSetListener() {
            @Override
            public void onTimeSet(TimePicker view, int hourOfDay, int minute) {
TextView tv1= (TextView) findViewById(R.id.textView);
                String theTime=String.format("你选择的时间：%d:%d",hourOfDay,minute);
tv1.setText(theTime);
            }
        };
        //定义TimePickerDialog对话框
        TimePickerDialogtimePickerDialog=
                                newTimePickerDialog(MainActivity.this,onTimeSetListener,0,0,true);
        timePickerDialog.show();//显示对话框
    }
```

上述代码中，TimePickerDialog()构造函数的参数依次为当前活动上下文、OnTimeSetListener监听器、初始小时、初始分钟以及是否为 24 小时制，程序运行时显示的时间选取对话框如图 3-24 所示。

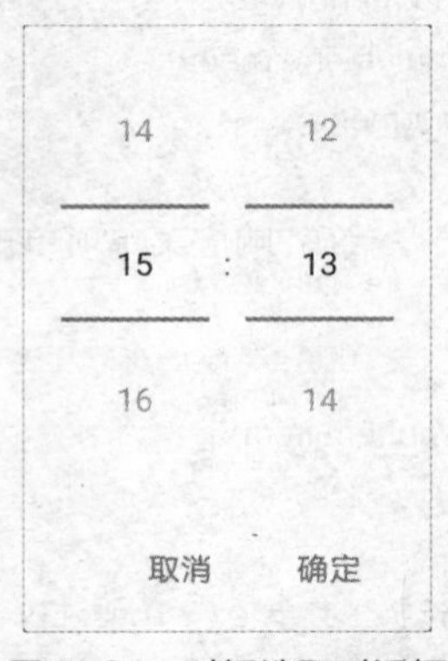

图 3-24　时间选取对话框

3.5　菜单

在手机和平板电脑等设备中，因为空间有限，所以不再像 PC 一样为应用程序配置菜单栏。Android 提供了一种隐藏的菜单，只在需要的时候展示出来。图 3-25 所示的活动标题栏右侧显示了一个三点符号，这就是菜单按钮，单击菜单按钮可展开活动的菜单。

图 3-25　带有菜单按钮的标题栏

为活动添加菜单的具体操作步骤如下所述。本节实例项目：源代码\03\LearnMenu。

（1）在 AndroidStudio 中创建一个新项目，将应用名称设置为 LearnMenu，并为项目添加一个空活动。

（2）在项目窗口中右击 res 文件夹，在弹出的快捷菜单中选择“New/Directory”命令，创建一个文件夹，文件夹命名为 menu。

（3）右击 menu 文件夹，在弹出的快捷菜单中选择“New/Menu resource file”命令，创建一个菜单资源文件，文件命名为 mainmenu。

（4）修改 mainmenu.xml，代码如下。

```
<?xml version="1.0" encoding="utf-8"?>
<menu xmlns:android="http://schemas.android.com/apk/res/android">
    <item android:id="@+id/item1" android:title="菜单项1"/>
    <item android:id="@+id/item2" android:title="菜单项2"/>
</menu>
```

其中，在<menu>标签中为菜单定义，每个<item>标签定义一个菜单项。

（5）返回 MainActivity.java 代码编辑窗口。在 Android Studio 菜单中选择“Code/Override Methods”命令，打开 SelectMethod to Override/Implement 对话框。在对话框中选中 onCreateOptionsMenu 和 onOptionsItemSelected 方法（按住【Ctrl】键可同时选中多个方法），最后单击“OK”按钮来添加两个方法的代码结构。

（6）编写 onCreateOptionsMenu 方法代码，为活动关联菜单，代码如下。

```
@Override
public boolean onCreateOptionsMenu(Menu menu) {
        getMenuInflater().inflate(R.menu.mainmenu,menu);
        return super.onCreateOptionsMenu(menu);
    }
```

（7）编写 onOptionsItemSelected 方法代码，响应菜单项单击事件，在用户选择菜单项时显示 Toast 通知，代码如下。

```
@Override
public boolean onOptionsItemSelected(MenuItem item) {
        switch (item.getItemId()){
            case R.id.item1:
                Toast.makeText(this, "你选择了菜单项1", Toast.LENGTH_SHORT).show();
                break;
            case R.id.item2:
                Toast.makeText(this, "你选择了菜单项2", Toast.LENGTH_SHORT).show();
        }
        return super.onOptionsItemSelected(item);
    }
```

（8）运行项目，测试运行效果。程序运行时显示的菜单如图 3-26 所示。

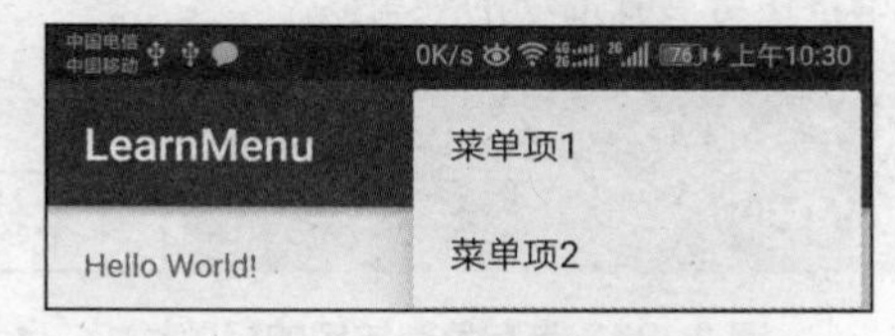

图 3-26　活动展示的菜单

选择菜单中的命令，在屏幕下方可显示对应的 Toast 通知，如图 3-27 所示。

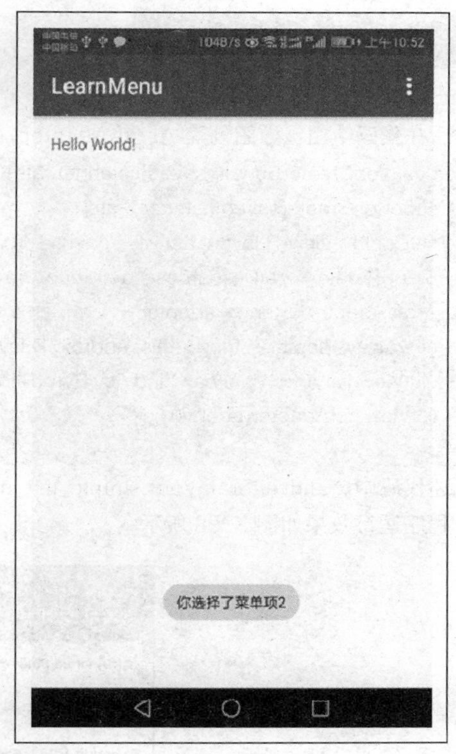

图 3-27 选择命令时显示的 Toast 通知

3.6 ListView

ListView 控件用于创建列表。比如联系人、QQ 聊天记录等，都可用列表来实现。ListView 允许用户通过上下滑动的方式将屏幕之外的内容滚动到屏幕内。

3.6.1 ListView 简单用法

为布局添加 ListView 控件非常简单，代码如下。本节实例项目：源代码\03\LearnListView。

Android 用户界面之常用控件 ListView

```
<ListView
        android:layout_width="match_parent"
        android:layout_height="match_parent"
        android:id="@+id/listView"/>
```

这样的 ListView 还没有数据可显示，可调用 setAdapter()方法为 ListView 绑定数据适配器。如果使用字符串数组作为 ListView 数据源，则可使用 ArrayAdapter。

首先在字符串资源文件 string.xml 中定义字符串数组，例如如下代码。

```
<resources>
    ...
    <string-array name="LearnListViewData">
        <item>使用Android Studio环境</item>
        <item>Android Studio实战</item>
```

```
            <item>Android编程权威指南</item>
        </string-array>
</resources>
```

然后在代码中用该数组创建 ArrayAdapter，并绑定到 ListView，例如如下代码。

```
protected void onCreate(Bundle savedInstanceState) {
        super.onCreate(savedInstanceState);
        setContentView(R.layout.activity_main);
        String[] data = getResources().getStringArray(R.array.LearnListViewData);
        ArrayAdapter<String> adapter =
        newArrayAdapter<String>(this, android.R.layout.simple_list_item_1, data);
        ListView listView=(ListView)findViewById(R.id.listView);
        listView.setAdapter(adapter);
}
```

上述代码中，android.R.layout.simple_list_item_1 是 Android 内置的列表项样式，用于显示简单文本，程序运行效果如图 3-28 所示。

图 3-28　简单的 ListView

3.6.2　自定义 ListView 列表项布局

普通 ListView 的每个列表项只显示一段文本，通过自定义，可以让列表项显示更丰富的内容。

通过自定义，在列表项中显示图书封面图片和书名的具体操作步骤如下。本节实例项目：源代码\03\LearnListView2。

（1）在 AndroidStudio 中创建一个新项目，将应用名称设置为 LearnListView2，并为项目添加一个空活动。

（2）修改 activity_main.xml，为主活动布局添加一个 ListView 控件。

（3）修改字符串资源文件 res\values\string.xml，添加图书书名数组定义。

（4）将图书封面图片复制到 res\drawable 文件夹中。

（5）在 res\layout 文件夹中添加 ListView 列表项布局文件 book_item.xml，代码如下。

```
<?xml version="1.0" encoding="utf-8"?>
<LinearLayout xmlns:android="http://schemas.android.com/apk/res/android"
    xmlns:app="http://schemas.android.com/apk/res-auto"
    android:orientation="horizontal" android:layout_width="match_parent"
    android:layout_height="wrap_content ">
    <ImageView
        android:layout_width="wrap_content"
        android:layout_height="wrap_content"
        android:id="@+id/bookpic"/>
    <TextView
        android:layout_gravity="center_vertical"
        android:layout_width="wrap_content"
        android:layout_height="wrap_content"
        android:id="@+id/bookname" />
</LinearLayout>
```

（6）在 MainActivity.java 相同的文件夹中添加一个 Java 类，命名为 Book.java，代码如下。

```
package com.example.xbg.learnlistview2;
public class Book {
    private String picId;
    private String name;
    public Book(String picId, String name) {
        this.picId = picId;
        this.name = name;
    }
    public String getName() { return name; }
    public String getPicId() { return picId;   }
}
```

（7）在 MainActivity.java 相同的文件夹中添加一个自定义的适配器 Java 类，命名为 BookAdapter.java，代码如下。

```
package com.example.xbg.learnlistview2;
import android.content.Context;
…
public class BookAdapter extends ArrayAdapter<Book> {
    public int resId;
    public BookAdapter(Context context, int resource, List<Book> objects) {
        super(context, resource, objects);
        resId=resource;
    }
    @Override
    public View getView(int position, View convertView, ViewGroup parent) {
        Book book=getItem(position);//获得当前列表项Book对象
        View view= LayoutInflater.from(getContext()).inflate(resId,parent,false);
        TextView bookname=(TextView)view.findViewById(R.id.bookname);
        bookname.setText(book.getName());
```

```
            ImageView bookpic=(ImageView)view.findViewById(R.id.bookpic);
            bookpic.setImageResource(book.getPicId());
            return view;
        }
}
```

（8）修改 MainActivity.java，使用自定义的适配器填充 ListView，代码如下。

```
package com.example.xbg.learnlistview2;
import android.support.v7.app.AppCompatActivity;
...
public class MainActivity extends AppCompatActivity {
    @Override
    protected void onCreate(Bundle savedInstanceState) {
        super.onCreate(savedInstanceState);
        setContentView(R.layout.activity_main);
        //创建图书列表
        List<Book> bookList=new ArrayList<Book>();
        String[] names = getResources().getStringArray(R.array.LearnListViewData);
        Book book1=new Book(R.drawable.pic1,names[0]);
        bookList.add(book1);
        Book book2=new Book(R.drawable.pic2,names[1]);
        bookList.add(book2);
        Book book3=new Book(R.drawable.pic3,names[2]);
        bookList.add(book3);
        BookAdapter adapter=new BookAdapter(MainActivity.this,R.layout.book_item,bookList);
        ListView listView=(ListView)findViewById(R.id.listView);
        listView.setAdapter(adapter);
    }
}
```

（9）运行项目，测试运行效果。程序运行效果如图 3-29 所示。每个列表项都包含图书封面图片和书名。

图 3-29　自定义 ListView 列表项布局示例

3.6.3 处理 ListView 单击事件

要使 ListView 响应用户单击事件，需要调用 setOnItemClickListener() 方法绑定 OnItemClickListener 监听器。例如，修改 3.6.2 节实例中的 MainActivity.java 为 ListView 绑定监听器的代码如下。

```
protected void onCreate(Bundle savedInstanceState) {
...
        ListView listView=(ListView)findViewById(R.id.listView);
        listView.setAdapter(adapter);
        listView.setOnItemClickListener(new AdapterView.OnItemClickListener() {
            @Override
            public void onItemClick(AdapterView<?> parent, View view, int position, long id) {
                Book book=bookList.get(position);
                Toast.makeText(MainActivity.this,book.getName(),Toast.LENGTH_LONG).show();
            }
        });
    }
```

重新运行程序，并单击一个列表项，即可在 Toast 通知中显示图书名称，如图 3-30 所示。

图 3-30　ListView 响应单击事件

3.7 RecyclerView

ListView 在过去的 Android 应用中发挥了巨大的作用，其功能强大，但缺点明显。ListView 只能实现数据的垂直滚动，而且在不采取措施优化时性能极差。Android 提供了另一个功能更强、效率更

使用 RecyclerView

高的滚动控件——RecyclerView。RecyclerView 可看作是升级版的 ListView，Android 官方推荐使用 RecyclerView 来实现滚动列表。

3.7.1 RecyclerView 基本用法

要使用 RecyclerView 控件，首先需要在 app\buil.gradle 文件的 dependencies 闭包中添加支持库，例如如下代码。

```
dependencies {
    ...
    compile 'com.android.support:recyclerview-v7:25.3.1'
}
```

在 Android Studio 中打开布局文件的 Design 视图，在 Pallette 窗口的控件列表中第一次使用 RecyclerView 控件时，Android Studio 会提示添加支持库，需要手动添加。

在布局文件中，可用如下代码添加 RecyclerView 控件。

```
<android.support.v7.widget.RecyclerView
        android:layout_width="match_parent"
        android:layout_height="match_parent"
        android:id="@+id/recyclerView"/>
```

使用 RecyclerView 控件的具体操作步骤如下。本节实例项目：源代码\03\LearnRecyclerView。

（1）在 AndroidStudio 中创建一个新项目，将应用名称设置为 LearnRecyclerView，并为项目添加一个空活动。

（2）修改 activity_main.xml，为主活动布局添加一个 RecyclerView 控件。

（3）修改字符串资源文件 res\values\string.xml，添加图书书名数组定义。

（4）修改 MainActivity.java，添加 RecyclerView 相关处理代码，代码如下。

```
package com.example.xbg.learnrecyclerview;
import android.support.v7.app.AppCompatActivity;
...
public class MainActivity extends AppCompatActivity {
    private String[] bookMames;//用于引用字符串资源文件中的字符串数组
    @Override
    protected void onCreate(Bundle savedInstanceState) {
        super.onCreate(savedInstanceState);
        setContentView(R.layout.activity_main);
        //获得字符串资源文件中的字符串数组
        bookMames = getResources().getStringArray(R.array.LearnListViewData);
        //获得布局中的RecyclerView控件
        RecyclerView recyclerView=(RecyclerView)findViewById(R.id.recyclerView);
        //设置RecyclerView控件布局管理器类型
        recyclerView.setLayoutManager(new LinearLayoutManager(this));
        //定义RecyclerView适配器
        recyclerView.setAdapter(new RecyclerView.Adapter() {
            //定义用于处理每个列表项的ViewHolder，每个ViewHolder包含了列表项的各个控件
            class ItemViewHolder extends RecyclerView.ViewHolder{
```

```
            private TextView itemTextView;
            public ItemViewHolder(TextView itemView) {
                super(itemView);
                itemTextView=itemView;
            }
            public TextView getItemTextView() {
            return itemTextView;
            }
        }
        @Override
        public RecyclerView.ViewHolder onCreateViewHolder(ViewGroup parent, int viewType) {
            //生成列表项ViewHolder
            return new ItemViewHolder(new TextView(parent.getContext()));
        }
        @Override
        public void onBindViewHolder(RecyclerView.ViewHolder holder, int position) {
            //为列表项设置数据
            ItemViewHolder itemViewHolder=(ItemViewHolder)holder;
            itemViewHolder.getItemTextView().setText(position+":"+bookMames[position%3]);
        }
        @Override
        public int getItemCount() {
            return 50;//指定RecyclerView列表项的数目
            }
    });
  }
}
```

（5）运行项目，测试运行效果。程序运行效果如图 3-31 所示，超出屏幕的数据可通过上下滚动展示出来。

图 3-31　自定义的 RecyclerView 界面

使用资源文件自定义列表项

3.7.2 自定义 RecyclerView 列表项布局

与 ListView 类似，用户可以自定义 RecyclerView 列表项的布局。

在如下实例中，通过自定义，在 RecyclerView 列表项中显示图书封面图片和书名。本节实例项目：源代码\03\LearnRecyclerView2。

（1）在 AndroidStudio 中创建一个新项目，将应用名称设置为 LearnRecyclerView2，并为项目添加一个空活动。

（2）修改 activity_main.xml，在主活动布局中添加一个 RecyclerView 控件（与项目 LearnListView2 相同）。

（3）修改字符串资源文件 res\values\string.xml，添加图书书名数组定义（与项目 LearnListView2 相同）。

（4）将图书封面图片复制到 res\drawable 文件夹中。

（5）在 res\layout 文件夹中添加 ListView 列表项布局文件 book_item.xml（与项目 LearnListView2 相同）。

（6）在 MainActivity.java 相同的文件夹中添加一个 Java 类，命名为 Book.java（与项目 LearnListView2 相同）。

（7）在 MainActivity.java 相同的文件夹中添加一个自定义的适配器 Java 类，命名为 BookAdapter.java，代码如下。

```
package com.example.xbg.learnrecyclerview2;
import android.content.Context;
...
public class BookAdapter extends RecyclerView.Adapter<BookAdapter.ItemViewHolder> {
    private List<Book> bookList;
    static class ItemViewHolder extends RecyclerView.ViewHolder{
        private TextView bookName;
        private ImageView bookPic;
        public ItemViewHolder(View itemView) {
            super(itemView);
            bookName=(TextView) itemView.findViewById(R.id.bookname);
            bookPic=(ImageView) itemView.findViewById(R.id.bookpic);
        }
    }
    public BookAdapter(List<Book> bookList) {
        this.bookList = bookList;
    }
    @Override
    public ItemViewHolder onCreateViewHolder(ViewGroup parent, int viewType) {
        View view=LayoutInflater.from(parent.getContext())
                .inflate(R.layout.book_item,parent,false);
        ItemViewHolder itemViewHolder=new ItemViewHolder(view);
        return itemViewHolder;
    }
    @Override
    public void onBindViewHolder(ItemViewHolder holder, int position) {
```

```
            Book book=bookList.get(position);
            holder.bookName.setText(book.getName());
            holder.bookPic.setImageResource(book.getPicId());
    }
    @Override
    public int getItemCount() {
            return bookList.size();
    }
}
```

BookAdapter类实现的适配器用于填充RecyclerView列表，其每个列表项包含一个ImageView和一个TextView。BookAdapter类继承自RecyclerView.Adapter，其泛型指定为BookAdapter.ItemViewHolder。ItemViewHolder是BookAdapter类中定义的一个内部类。ItemViewHolder继承自RecyclerView.ViewHolder，其构造函数的参数是一个View对象，通常为RecyclerView列表项的布局。所以可调用findViewById()方法来获取布局中的控件。

BookAdapter类继承自RecyclerView.Adapter，所以必须重写onCreateViewHolder()、onBindViewHolder()和getItemCount()方法。

（8）修改MainActivity.java，使用自定义的适配器填充RecyclerView，代码如下。

```
package com.example.xbg.learnrecyclerview2;
import android.support.v7.app.AppCompatActivity;
...
public class MainActivity extends AppCompatActivity {
    private String[] bookMames;//用于引用字符串资源文件中的字符串数组
    private List<Book> bookList=new ArrayList<Book>();//用于保存图书信息
    @Override
    protected void onCreate(Bundle savedInstanceState) {
        super.onCreate(savedInstanceState);
        setContentView(R.layout.activity_main);
        //获得字符串资源文件中的字符串数组
        bookMames = getResources().getStringArray(R.array.LearnListViewData);
        //初始化图书信息，这里只是示例，所以使用了重复的数据
        for (int i=0;i<20;i++){
            Book book1=new Book(R.drawable.pic1,bookMames[0]);
            bookList.add(book1);
            Book book2=new Book(R.drawable.pic2,bookMames[1]);
            bookList.add(book2);
            Book book3=new Book(R.drawable.pic3,bookMames[2]);
            bookList.add(book3);
        }
        //获得布局中的RecyclerView控件
        RecyclerView recyclerView=(RecyclerView)findViewById(R.id.recyclerView);
        //设置RecyclerView控件布局管理器类型
        recyclerView.setLayoutManager(new LinearLayoutManager(this));
        //定义RecyclerView适配器
        recyclerView.setAdapter(new BookAdapter(bookList) );
    }
```

```
}
```

（9）运行项目，测试运行效果。程序运行效果如图 3-32 所示。

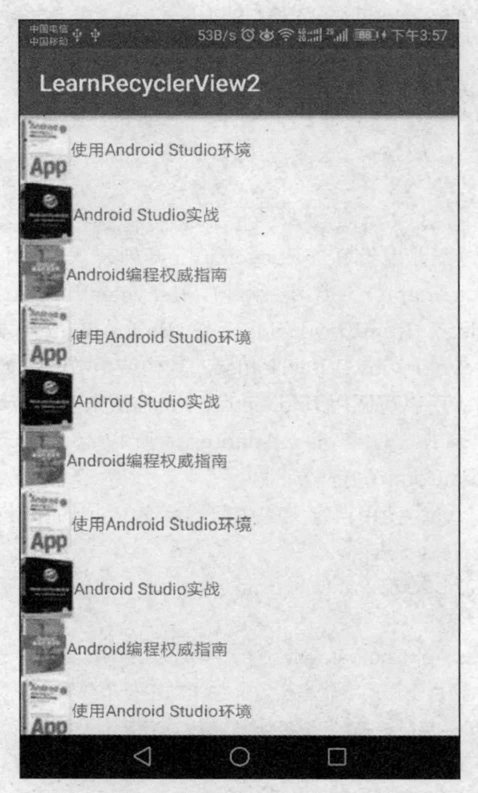

图 3-32　自定义 RecyclerView 列表项布局

RecyclerView 的布局样式

3.7.3　RecyclerView 布局

通过 setLayoutManager() 方法可以设置 RecyclerView 布局类型，RecyclerView 支持 LinearLayoutManager（线性布局）、StaggeredGridLayoutManager（瀑布流布局）和 GridLayoutManager（网格布局）等布局。

3.7.2 小节的 LearnRecyclerView2 实例使用 LinearLayoutManager 实现了垂直滚动列表，稍加修改，可以实现水平滚动列表。

首先，修改列表项布局文件 book_item.xml，代码如下。

```
<?xml version="1.0" encoding="utf-8"?>
<LinearLayoutxmlns:android="http://schemas.android.com/apk/res/android"
    xmlns:app="http://schemas.android.com/apk/res-auto"
    android:orientation="vertical" android:layout_width="100dp"
    android:layout_height="wrap_content">
    <ImageView
        android:layout_width="wrap_content"
        android:layout_height="wrap_content"
        android:layout_gravity="center_vertical"
        android:id="@+id/bookpic"/>
```

```
    <TextView
        android:layout_gravity="center_vertical"
        android:layout_width="wrap_content"
        android:layout_height="wrap_content"
        android:id="@+id/bookname" />
</LinearLayout>
```

将 LinearLayout 布局方向设置为垂直，即布局内部的控件按垂直方向排列。将布局宽度设置为 100dp，这是为了避免文本控件中的文字过长而导致一个列表项占用太多横向空间。

然后在 MainActivity.java 中修改 setLayoutManager()方法调用语句，代码如下。

```
recyclerView.setLayoutManager(new LinearLayoutManager(this,
                                                LinearLayoutManager.HORIZONTAL,false));
```

LinearLayoutManager 构造函数的第 1 个参数为当前活动上下文，第 2 个参数为布局方向，第 3 个参数 false 表示布局不反转（true 表示反转，反转指控件按从右到左的顺序排列）。

修改后的程序运行效果如图 3-33 所示。

图 3-33　水平排列的 RecyclerView

在 MainActivity.java 中修改 setLayoutManager()方法调用语句，使用瀑布流布局，代码如下。

```
StaggeredGridLayoutManager layoutManager=
                  new StaggeredGridLayoutManager(3,StaggeredGridLayoutManager.VERTICAL);
recyclerView.setLayoutManager(layoutManager);
```

StaggeredGridLayoutManager 构造函数的第 2 个参数为布局方向（垂直），此时第 1 个参数表示垂直方向上的列表项分 3 列。如果方向为水平，则表示列表项水平分 3 行。

修改后的程序运行效果如图 3-34 所示。

图 3-34　3 列瀑布布局的 RecyclerView

3.7.4　处理 RecyclerView 单击事件

与 ListView 不同，RecyclerView 没有提供类似的 setOnItemClickListener()方法来设置列表项监听器，但可通过 RecyclerView 列表项中的各个控件来设置。

修改 3.7.2 小节 LearnRecyclerView2 实例中的 BookAdapter.java，为列表项控件设置监听器，代码如下。

```
publicItemViewHolderonCreateViewHolder(ViewGroup parent, int viewType) {
        View view=LayoutInflater.from(parent.getContext())
                .inflate(R.layout.book_item,parent,false);
        finalItemViewHolderitemViewHolder=new ItemViewHolder(view);
        itemViewHolder.bookPic.setOnClickListener(new View.OnClickListener() {
            @Override
            public void onClick(View v) {
                int index=itemViewHolder.getAdapterPosition();
                Book book=bookList.get(index);
                Toast.makeText(v.getContext(),
                        "你单击了："+book.getName()+" 的图片",Toast.LENGTH_LONG).show();
            }
        });
        itemViewHolder.bookName.setOnClickListener(new View.OnClickListener() {
            @Override
            public void onClick(View v) {
                int index=itemViewHolder.getAdapterPosition();
                Book book=bookList.get(index);
                Toast.makeText(v.getContext(),
                        "你单击了："+book.getName()+" 的书名",Toast.LENGTH_LONG).show();
```

```
        }
    });
    returnitemViewHolder;
}
```

修改后的程序运行效果如图 3-35 所示。

图 3-35　处理 RecyclerView 单击事件

3.8　编程实践：用户登录界面设计

本节将综合应用本章所学知识，设计一个用户登录界面，效果如图 3-36 所示。如果输入的用户名和密码均正确，则显示登录成功信息，如图 3-37 所示；否则显示错误提示，如图 3-38 所示。

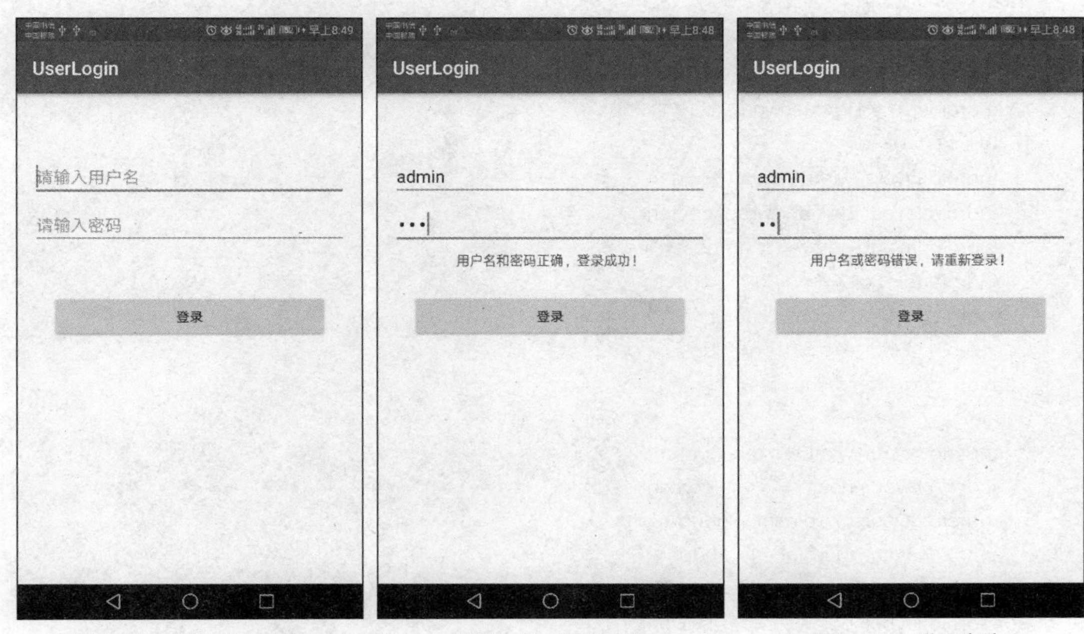

图 3-36　用户登录界面　　图 3-37　登录成功　　图 3-38　登录失败

具体操作步骤如下。

（1）在 AndroidStudio 中创建一个新项目，将应用名称设置为 UserLogin，并为项目添加一个空活动。

（2）修改 activity_main.xml，在主活动布局中添加控件，代码如下。

```
<?xml version="1.0" encoding="utf-8"?>
<RelativeLayout xmlns:android="http://schemas.android.com/apk/res/android"
    xmlns:tools="http://schemas.android.com/tools"
    android:id="@+id/activity_main"
    android:layout_width="match_parent"
    android:layout_height="match_parent"
    android:paddingBottom="@dimen/activity_vertical_margin"
    android:paddingLeft="@dimen/activity_horizontal_margin"
    android:paddingRight="@dimen/activity_horizontal_margin"
    android:paddingTop="@dimen/activity_vertical_margin"
    tools:context="com.example.xbg.userlogin.MainActivity">
    <EditText
        android:layout_width="match_parent"
        android:layout_height="wrap_content"
        android:inputType="textPersonName"
        android:hint="请输入用户名"
        android:layout_marginTop="50dp"
        android:id="@+id/userName" />
    <EditText
        android:layout_width="match_parent"
        android:layout_height="wrap_content"
        android:inputType="textPassword"
        android:layout_marginTop="5dp"
        android:layout_below="@+id/userName"
        android:hint="请输入密码"
        android:id="@+id/password" />
    <TextView
        android:layout_width="match_parent"
        android:layout_height="wrap_content"
        android:layout_below="@+id/password"
        android:textColor="#FF0000"
        android:textAlignment="center"
        android:layout_marginTop="5dp"
        android:id="@+id/showMsg" />
    <Button
        android:layout_width="match_parent"
        android:layout_height="wrap_content"
        android:layout_marginTop="50dp"
        android:layout_marginLeft="20dp"
        android:layout_marginRight="20dp"
        android:layout_below="@id/password"
        android:onClick="checkLog"
```

```
        android:text="登录" />
</RelativeLayout>
```

（3）修改字符串资源文件 res\values/string.xml，添加用于验证用户登录信息的用户名和对应的密码，代码如下。

```
<resources>
    <string name="app_name">UserLogin</string>
    <string-array name="UserNames">
        <item>admin</item>
        <item>Administrator</item>
        <item>Android</item>
    </string-array>
    <string-array name="Passwords">
        <item>123</item>
        <item>123456</item>
        <item>111111</item>
    </string-array>
</resources>
```

（4）修改 MainActivity.java，代码如下。

```
package com.example.xbg.userlogin;
import android.support.v7.app.AppCompatActivity;
import android.os.Bundle;
import android.text.Editable;
import android.text.TextWatcher;
import android.view.View;
import android.widget.EditText;
import android.widget.TextView;
public class MainActivity extends AppCompatActivity {
    private String[] UserNames;
    private String[] Passwords;
    EditText etUserName;
    EditText etPassword;
    TextView tvShow;
    @Override
    protected void onCreate(Bundle savedInstanceState) {
        super.onCreate(savedInstanceState);
        setContentView(R.layout.activity_main);
        UserNames= getResources().getStringArray(R.array.UserNames);
        Passwords= getResources().getStringArray(R.array.Passwords);
        tvShow=(TextView)findViewById(R.id.showMsg);
        etUserName=(EditText)findViewById(R.id.userName);
        etUserName.addTextChangedListener(new TextWatcher() {
            @Override
            public void beforeTextChanged(CharSequence s, int start, int count, int after) {
                //文本改变之前调用
            }
```

```
            @Override
            public void onTextChanged(CharSequence s, int start, int before, int count) {
                //文本正在改变时调用
                tvShow.setText("");//在输入新的用户名时，清除TextView显示的提示信息
            }
            @Override
            public void afterTextChanged(Editable s) {
                //文本改变之后调用
            }
        });
        etPassword=(EditText)findViewById(R.id.password);
        etPassword.addTextChangedListener(new TextWatcher() {
            @Override
            public void beforeTextChanged(CharSequence s, int start, int count, int after) {}
            @Override
            public void onTextChanged(CharSequence s, int start, int before, int count) {
                tvShow.setText("");//在输入新的用户名时，清除TextView显示的提示信息
            }
            @Override
            public void afterTextChanged(Editable s) {}
        });
    }
    public void checkLog(View view){
        String userName=etUserName.getText().toString();//获得收入的用户名
        String password=etPassword.getText().toString();//获得收入的密码
        //判断用户名是否正确
        int index;
        boolean isLoged=false;
        for(index=0;index<UserNames.length;index++){
            if(userName.equals(UserNames[index])) {
                if (password.equals(Passwords[index])) {
                    isLoged = true;
                    break;//用户名和密码都正确，不再继续验证
                }
            }
        }
        if(isLoged){
            //执行登录成功操作
            tvShow.setText("用户名和密码正确，登录成功！");
        }else {
            //执行登录失败操作
            tvShow.setText("用户名或密码错误，请重新登录！");
        }
    }
}
```

当用户修改用户名或密码时，需清除上一次的验证信息，所以需要为用户名和密码输入控件添加文本改变监听器。addTextChangedListener()方法用于为 EditView 控件添加文本改变监听器，其参数为 TextWatcher 对象，需要为其重写 beforeTextChanged()、onTextChanged()和 onTextChanged()方法。本例只需要在 onTextChanged()中清除验证信息即可，另外两个方法仍然需要，只是不需要添加代码。

（5）运行项目，测试运行效果。

3.9 小结

本章主要讲解了和界面设计有关的基础知识。首先讲解了布局，布局就是用户界面。最常用的布局是 LinearLeyout 和 RelativeLayout。Android 也提供了其他的多种布局供用户选择。布局文件是一个 XML 文件，可使 UI 设计与逻辑代码分离，体现了现代应用程序设计的理念。

另外介绍了通用 UI 组件常用的控件。Android 在 android.widget 包中提供了基础控件，有兴趣的读者可在帮助文档中查看 android.widget 包中包含的控件。本章最后还重点讲解了 ListView 和 RecyclerView 这两个高级控件，从功能上两者区别很小，但 RecyclerView 效率更高。

另外，本章还讲解了消息通知、对话框和菜单，这些都是应用程序除界面控件之外和用户交互的手段。

3.10 习题

1. 简述视图（View）和视图组（ViewGroup）的联系和区别。
2. 列举 5 种基本的 Android 控件。
3. 简述如何使用 Toast。
4. 简述如何使用 Notification。
5. 简述如何使用 RecyclerView。
6. 修改本章 3.8 节的实例，改为使用对话框显示登录验证信息。

PART04

第4章

广播机制

重点知识：

- 静态注册广播接收器
- 动态注册和注销广播接收器
- 接收系统广播
- 发送本地广播
- 广播接收器优先级与有序广播

■ 广播是一种群发机制。就像学校的广播，可以让学校的所有人员接收到重要通知。在网络通信中，如果使用广播地址，数据包就会发送到网络中的所有计算机。

Android 引入了类似的广播机制来实现应用内部、应用之间的消息通知。本章将学习如何在应用中发送和接收广播。

4.1 广播机制简介

Android 的广播机制非常灵活。广播可来自于系统，也可来自其他应用，甚至于来自应用内部的其他模块。应用程序可以只对感兴趣的广播进行注册，也只有注册了的广播才可以接收到。

Android 中的广播可分为标准广播和有序广播两种类型。

- 标准广播：标准广播在发出后，所有接收器均可接收到广播消息，各个接收器之间没有先后顺序之分。标准广播发出后，不可能被中断。
- 有序广播：有序广播在发出后，同一时间只有优先级较高的一个接收器才能接收到广播消息。只有在优先级较高的接收器处理完广播消息后，广播才能继续向优先级较低的接收器继续传递。在当前接收器中可中断广播，使后继优先级较低的接收器无法收到广播消息。

Android 提供了一套完整的 API，用于发送和接收广播。发送广播时，类似于 Activity，使用 Intent 对象来传递数据；接收广播时，使用广播接收器（BroadcastReceiver）。

4.2 使用广播接收器

Android 提供了一个 BroadcastReceiver 类。通过扩展该类，并重写 onReceive()方法，即可创建一个广播接收器。广播接收器接收到广播消息时，执行 onReceive()方法。广播接收器必须进行注册，有静态注册和动态注册两种注册方法。

4.2.1 静态注册广播接收器

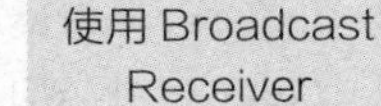

使用 Broadcast Receiver

静态注册是指在应用程序的清单文件 AndroidManifest.xml 中添加广播接收器的注册信息。

通过静态注册的方式使用广播接收器的具体操作步骤如下。实例项目：源代码\04\LearnBroadcastReceiver。

（1）创建一个新项目，将应用名称设置为 LearnBroadcastReceiver，并为项目添加一个空活动。

（2）在项目窗口中右击 MainActivity.java 所在的文件夹，在弹出的快捷菜单中选择“New/Other/Broadcast Receiver”命令，创建一个 Java 类，命名为 MyReceiver。

在通过菜单命令创建广播接收器类时，Android Studio 自动在 AndroidManifest.xml 完成注册。

（3）编写 MyReceiver.java 代码，实现广播接收器，在收到广播消息时弹出一个 Toast 通知，代码如下。

```
package com.example.xbg.learnbroadcastreceiver;
import android.content.BroadcastReceiver;
import android.content.Context;
import android.content.Intent;
import android.widget.Toast;
public class MyReceiver extends BroadcastReceiver {
    public MyReceiver() {
    }
    @Override
    public void onReceive(Context context, Intent intent) {
        Toast.makeText(context,"收到一个广播消息",Toast.LENGTH_LONG).show();
```

```
    }
}
```

（4）修改 activity_main.xml，在主活动布局中添加一个按钮控件，代码如下。

```
<Button
        android:text="发送广播"
        android:layout_width="match_parent"
        android:layout_height="wrap_content"
        android:onClick="sendMsg"
        android:id="@+id/button" />
```

（5）修改 MainActivity.java，实现用于发送广播消息的 sendMsg()方法，代码如下。

```
packagecom.example.xbg.learnbroadcastreceiver;
importandroid.content.Intent;
import android.support.v7.app.AppCompatActivity;
importandroid.os.Bundle;
importandroid.view.View;
public class MainActivity extends AppCompatActivity {
    @Override
    protected void onCreate(Bundle savedInstanceState) {
        super.onCreate(savedInstanceState);
        setContentView(R.layout.activity_main);
    }
    public void sendMsg(View view){
        sendBroadcast(new Intent(this,MyReceiver.class));
    }
}
```

sendBroadcast()方法用于发送一个广播消息，其参数为一个 Intent 对象。与在 Activity 中类似，在 Intent 中可添加各种数据。广播接收器的 onReceive()方法的第 2 个参数就是接收到的 Intent 对象，从中可取出数据。具体如何使用 Intent 对象传递数据，可参考本书 2.4 节“在活动之间传递数据”。

（6）运行项目，测试运行效果。

程序运行时，单击 发送广播 按钮即可发送广播消息，界面下方会显示 Toast 通知，如图 4-1 所示。

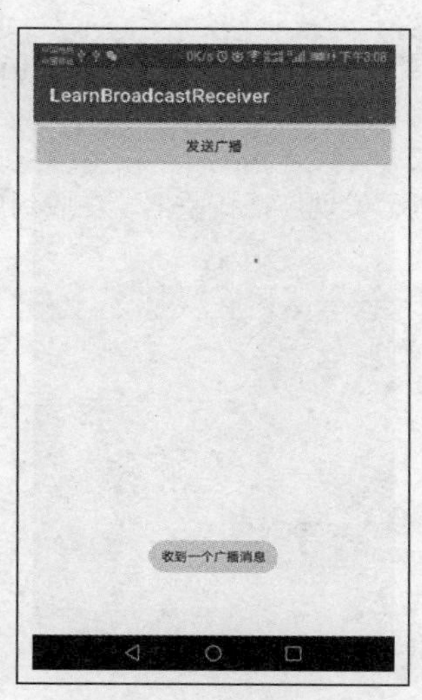

图 4-1 发送广播消息

4.2.2 动态注册和注销广播接收器

动态注册和注销
BroadcastReceiver

动态注册和注销广播接收器是指通过执行持续代码来注册和注销广播接收器，从而由用户控制是否启用接收器来接收广播。

动态注册和注销广播接收器的具体操作步骤如下。实例项目：源代码\04\LearnBroadcastReceiver2。

（1）创建一个新项目，将应用名称设置为 LearnBroadcastReceiver2，并为项目添加一个空活动。

（2）在项目窗口中右击 MainActivity.java 所在的文件夹，在弹出的快捷菜单中选择“New/Other/Java Class”命令，创建一个 Java 类，命名为 MyReceiver2，其父类为 android.content.BroadcastReceiver。

（3）编写 MyReceiver2.java 代码，实现广播接收器，在收到广播消息时弹出一个 Toast 通知，代码如下。

```
package com.example.xbg.learnbroadcastreceiver2;
importandroid.content.BroadcastReceiver;
importandroid.content.Context;
importandroid.content.Intent;
importandroid.widget.Toast;
public class MyReceiver2 extends BroadcastReceiver {
    public static String ACTION="learnbroadcastreceiver2.MyReceiver2";//定义操作
    public MyReceiver2() {
    }
    @Override
    public void onReceive(Context context, Intent intent) {
        Toast.makeText(context,"收到一个广播消息",Toast.LENGTH_LONG).show();
    }
}
```

与上小节实例中的 MyReceiver 类相比，这里的 MyReceiver2 类多了一个公共常量 ACTION。这是因为在注册广播接收器时需要指定一个操作，本例中用这里定义的 ACTION 常量来作为操作名称。

（4）修改 activity_main.xml，在主活动布局中添加 3 个按钮控件，代码如下。

```
<Button android:id="@+id/btnReg"
        android:onClick="registerMyReceiver"
        android:layout_width="wrap_content"
        android:layout_height="wrap_content"
        android:text="注册广播接收器" />
    <Button android:id="@+id/btnUnReg"
        android:onClick="unRegisterMyReceiver"
        android:layout_width="wrap_content"
        android:layout_height="wrap_content"
        android:text="注销广播接收器" />
    <Button android:id="@+id/btnSendMsg"
        android:onClick="sendMsg"
        android:layout_width="wrap_content"
        android:layout_height="wrap_content"
```

```
        android:text="发送广播消息" />
```

（5）修改 MainActivity.java，实现注册广播接收器、注销广播接收器和发送广播消息的方法，代码如下。

```
package com.example.xbg.learnbroadcastreceiver2;
import android.content.Intent;
import android.content.IntentFilter;
import android.support.v7.app.AppCompatActivity;
import android.os.Bundle;
import android.view.View;
public class MainActivity extends AppCompatActivity {
    private MyReceiver2 receiver=null;
    @Override
    protected void onCreate(Bundle savedInstanceState) {
        super.onCreate(savedInstanceState);
        setContentView(R.layout.activity_main);
    }
    public void registerMyReceiver(View view){
        //注册广播接收器
        if(receiver==null){
            receiver=new MyReceiver2();
            registerReceiver(receiver,new IntentFilter(MyReceiver2.ACTION));
        }
    }
    public   void   unRegisterMyReceiver(View view){
        //注销广播接收器
        if(receiver!=null){
            unregisterReceiver(receiver);
            receiver=null;
        }
    }
    public void sendMsg(View view){
        sendBroadcast(new Intent(MyReceiver2.ACTION));//发送广播消息
    }
}
```

registerReceiver()方法用于注册一个广播接收器，其参数为一个广播接收器对象和 IntentFilter 对象。IntentFilter 对象用自定义的操作作为参数，这样该接收器就只能接收到包含同样操作的 Intent 广播消息。unregisterReceiver()方法用于注销指定的接收器。

要使 registerReceiver()方法注册的广播接收器接收到广播消息，必须在发送广播时用包含了同样操作的 Intent 作为参数。

本例中使用了自定义的字符串作为操作来发送广播消息，这就实现了发送自定义广播。在第 2 章中介绍过，只有 IntentFilter 匹配的 Activity 才响应包含相同操作的 Intent。同样，在广播机制中，也只有 IntentFilter 匹配的接收器才响应使用包含了相同操作的 Intent 作为参数的广播消息。

（6）运行项目，测试运行效果。

程序运行时，活动主界面如图 4-2 所示。单击**发送广播消息**按钮，可发现没有接收器响应；单击**注册广播接收器**按钮，然后单击**发送广播消息**按钮，可看到屏幕下方弹出 Toast 通知，如图 4-3 所示；单击**注销广播接收器**按钮注销接收器，再单击**发送广播消息**按钮，可发现没有接收器响应。

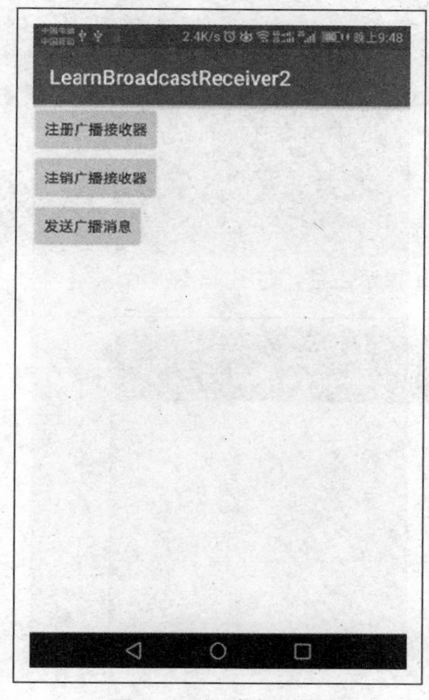

图 4-2 活动主界面

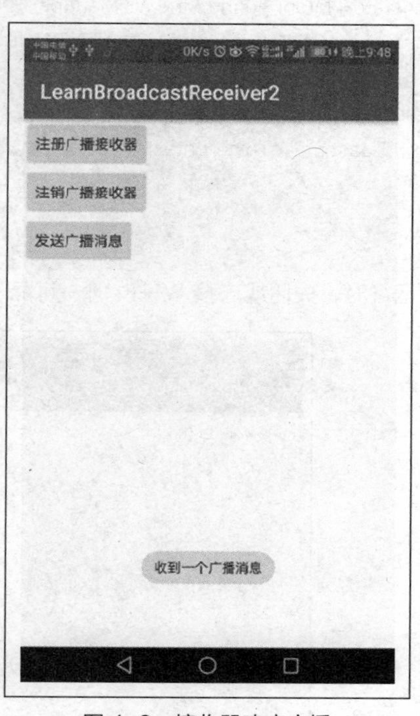

图 4-3 接收器响应广播

4.2.3 接收系统广播

Android 提供了一系列系统广播。在系统中发生某种事件时，系统就会自动发送相应的广播消息。例如，在系统 Wi-Fi 断开或连接时，系统会发送包含了 android.net.wifi.STATE_CHANGE 操作字符串的 Intent 广播消息。接收器接收到该消息时，可判定当前 Wi-Fi 连接是否可用。

在 Android SDK 安装目录下的 platforms\android-25\data 文件夹中的 broadcast_actions.txt 文件中，可查看对应 Android 版本支持的系统广播操作字符串。

要使接收器响应系统广播，需要在注册接收器时，在 IntentFilter 中指明可响应的操作。例如，要让接收器监听 Wi-Fi 连接状态变化，可在 AndroidManifest.xml 文件中用如下代码来注册接收器。实例项目：源代码\04\ReceiveSystemBroadcast。

```
<receiver android:name=".SysReceiver">
    <intent-filter>
        <actionandroid:name="android.net.wifi.Wi-Fi_STATE_CHANGED"/>
    </intent-filter>
</receiver>
```

自定义的广播接收器类 SysReceiver 的代码如下。

```
package com.example.xbg.receivesystembroadcast;
import android.content.BroadcastReceiver;
...
```

```
public class SysReceiver extends BroadcastReceiver {
    public SysReceiver() { }
    @Override
    public void onReceive(Context context, Intent intent) {
        int state= intent.getIntExtra(WifiManager.EXTRA_WIFI_STATE,0);
        if(state==WifiManager.WIFI_STATE_DISABLED){
            Toast.makeText(context,"Wi-FI连接已关闭！",Toast.LENGTH_SHORT).show();
        }else if(state==WifiManager.Wi-FI_STATE_ENABLED){
            Toast.makeText(context,"Wi-FI已连接！",Toast.LENGTH_SHORT).show();
        }
    }
}
```

程序运行时，关闭或连接 Wi-Fi 时，可显示 Toast 提示信息，如图 4-4 所示。

图 4-4　监听 Wi-Fi 连接状态

4.2.4　发送本地广播

当在活动中直接调用 sendBroadcast()方法发送广播时，默认为系统全局广播，可被其他应用中的接收器接收。如果不希望关键的广播消息被其他应用接收，则可使用本地广播，因为本地广播只能被当前应用中的接收器接收。

Android 提供了一个 LocalBroadcastManager（本地广播管理器）来管理本地广播的注册、注销和发送等操作。

如下实例代码实现了使用本地广播。实例项目：源代码\04\LocalBroadcast。

```
packagecom.example.xbg.localbroadcast;
importandroid.content.BroadcastReceiver;
...
public class MainActivity extends AppCompatActivity {
    privateMyReceiverlocalReceiver;
    privateLocalBroadcastManagerlocalBroadcastManager;
```

```
    @Override
    protected void onCreate(Bundle savedInstanceState) {
        super.onCreate(savedInstanceState);
        setContentView(R.layout.activity_main);
        //获得当前本地广播管理器
        localBroadcastManager= LocalBroadcastManager.getInstance(this);
        IntentFilterintentFilter=new IntentFilter("MyLocalBroadcastReceiver");
        localReceiver=new MyReceiver();//创建广播接收器对象
        localBroadcastManager.registerReceiver(localReceiver,intentFilter);//注册本地广播接收器
    }
    @Override
    protected void onDestroy() {
        super.onDestroy();
        localBroadcastManager.unregisterReceiver(localReceiver);//注销本地广播接收器
    }
    public void sendMyBroadcst(View view){
        Intent    intent=new Intent("MyLocalBroadcastReceiver");//用注册的操作创建Intent
        localBroadcastManager.sendBroadcast(intent);
    }


        public static class MyReceiver extends BroadcastReceiver {
        publicMyReceiver() { }
        @Override
        public void onReceive(Context context, Intent intent) {
            Toast.makeText(context,"收到一个本地广播消息", Toast.LENGTH_LONG).show();
        }
    }
}
```

程序运行时，单击 **发送广播消息** 按钮即可调用 sendMyBroadcst()方法发送本地广播，接收器会显示 Toast 消息，如图 4-5 所示。

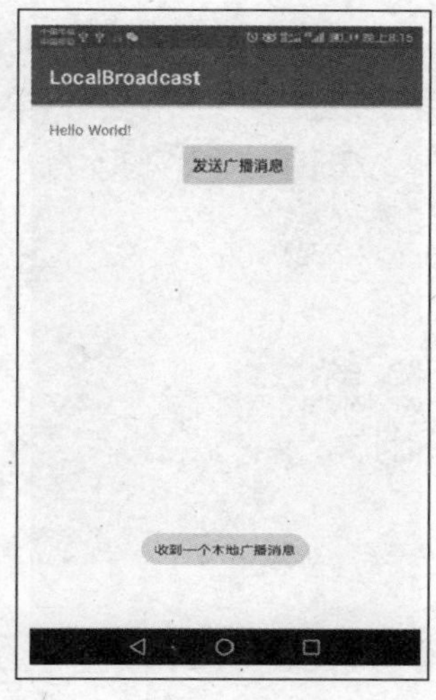

图 4-5　使用本地广播

4.3　广播接收器优先级与有序广播

BroadcastReceiver 的优先级

在前述案例中，对于同一个广播消息，接收器之间没有先后顺序之分，所有接收器同时接收到广播。在注册广播接收器时，可为接收器的 IntentFilter 设置优先级，优先级越高的接收器越先接收到广播。只有等优先级高的接收器处理完广播后，优先级较低的接收器才能接收到广播。在 AndroidManifest.xml 中静态注册接收器时，可使用<intent-filter>标签中的 android:priority 属性来设置广播接收器的优先级，例如如下代码。实例项目：源代码\04\PriorityOrderBroadcast。

```
<receiver android:name=".MyReceiver1" >
    <intent-filter android:priority="3">
        <action android:name="com.example.xbg.priorityorderbroadcast.ACTION"/>
    </intent-filter>
</receiver>
<receiver android:name=".MyReceiver2">
    <intent-filter android:priority="4">
        <action android:name="com.example.xbg.priorityorderbroadcast.ACTION"/>
    </intent-filter>
</receiver>
```

接收器类 MyReceiver1 的代码如下。

```
package com.example.xbg.priorityorderbroadcast;
import android.content.BroadcastReceiver;
...
public class MyReceiver1 extends BroadcastReceiver {
    public MyReceiver1() {}
    @Override
    public void onReceive(Context context, Intent intent) {
        System.out.println("MyReceiver1接收到广播！ ");
    }
}
```

接收器类 MyReceiver2 的代码如下。

```
package com.example.xbg.priorityorderbroadcast;
import android.content.BroadcastReceiver;
...
public class MyReceiver2 extends BroadcastReceiver {
    public MyReceiver2() {}
    @Override
    public void onReceive(Context context, Intent intent) {
        System.out.println("MyReceiver2接收到广播！ ");
    }
}
```

发送广播的代码如下。

```
public void sendMyBroadcast(View view){
    //发送广播
```

```
    Intent   intent=new Intent("com.example.xbg.priorityorderbroadcast.ACTION");
    sendBroadcast(intent);
}
```

在运行程序时，单击按钮即可调用 sendMyBroadcast()方法发送广播。在 Android Studio 监视应用运行情况的 Run 窗口中可查看广播接收器调用 System.out.println()方法输出的信息。改变 AndroidManifest.xml 文件中广播接收器的优先级，使 MyReceiver2 的优先级更高，可看到两个接收器输出的信息先后顺序也随之发生了变化，如图 4-6 所示。优先级高的接收器优先接收到了广播。

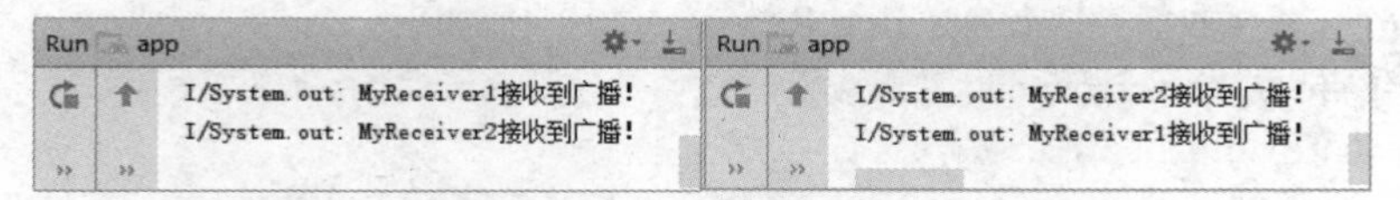

图 4-6　测试广播接收器优先级

上述实例说明，优先级决定了接收器接收广播的先后顺序，对于 sendBroadcast()方法发出的广播而言，都属于标准广播。

要发送有序广播，需调用 sendOrderedBroadcast()方法，例如如下代码。

```
sendOrderedBroadcast(intent,null);//发送有序广播
```

有序广播也由接收器的优先级决定接收的先后顺序。对于有序广播，在接收器的 onReceive()方法中调用 abortBroadcast()方法可终止广播传递，后续的接收器将无法接收到广播，例如如下代码。

```
public void onReceive(Context context, Intent intent) {
    System.out.println("MyReceiver1接收到广播！");
    abortBroadcast();//终止广播传递
}
```

4.4　编程实践：开机启动应用

本节将综合应用本章所介绍的知识，设计一个可以开机自动启动的应用，运行结果如图 4-7 所示。

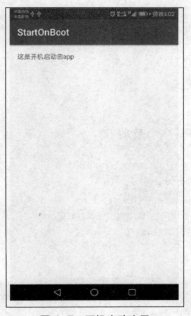

图 4-7　开机启动应用

系统开机启动后，会发送一个系统启动完成的全局系统广播，其字符串常量为android.intent.action.BOOT_COMPLETED。用广播接收器接收该广播，在onReceive()方法中启动应用程序，就可以实现开机启动应用，具体操作步骤如下。

（1）在AndroidStudio中创建一个新项目，将应用名称设置为StartOnBoot，并为项目添加一个空活动。

（2）修改activity_main.xml，将文本视图控件ID设置为textView。

（3）右击MainActivity.java所在的包，在弹出的快捷菜单中选择“New/Other/Broadcast Receiver”命令，创建一个广播接收器类，命名为BootReceiver，其代码如下。

```
package com.example.xbg.startonboot;
import android.content.BroadcastReceiver;
import android.content.Context;
import android.content.Intent;
import android.widget.Toast;
public class BootReceiver extends BroadcastReceiver {
    private static String BOOT_ACTION="android.intent.action.BOOT_COMPLETED";
    public BootReceiver() { }
    @Override
    public void onReceive(Context context, Intent intent) {
        Toast.makeText(context,"bootup",Toast.LENGTH_SHORT).show();
        String action=intent.getAction();
        if(action.equals(BOOT_ACTION) ){
            Intent boot=new Intent(context,MainActivity.class);//定义用于启动活动的Intent
            boot.addFlags(Intent.FLAG_ACTIVITY_NEW_TASK);
            context.startActivity(boot);//启动活动
        }
    }
}
```

（4）修改MainActivity.java，在启动活动时设置活动布局中的文本视图控件为显示文本，代码如下。

```
packagecom.example.xbg.startonboot;
import android.support.v7.app.AppCompatActivity;
importandroid.os.Bundle;
importandroid.widget.TextView;
public class MainActivity extends AppCompatActivity {
    @Override
    protected void onCreate(Bundle savedInstanceState) {
        super.onCreate(savedInstanceState);
        setContentView(R.layout.activity_main);
        TextViewtextView=(TextView)findViewById(R.id.textView);
        textView.setText("这是开机启动的app");
    }
}
```

（5）修改AndroidManifest.xml，添加权限和接收器IntentFilter，代码如下。

```
<?xml version="1.0" encoding="utf-8"?>
<manifest xmlns:android="http://schemas.android.com/apk/res/android"
```

```
    package="com.example.xbg.startonboot" >
    <uses-permissionandroid:name="android.permission.RECEIVE_BOOT_COMPLETED"/>
    <application
        android:allowBackup="true"
        android:icon="@mipmap/ic_launcher"
        android:label="@string/app_name"
        android:supportsRtl="true"
        android:theme="@style/AppTheme">
        <activity android:name=".MainActivity">
            <intent-filter>
                <action android:name="android.intent.action.MAIN" />
                <category android:name="android.intent.category.LAUNCHER" />
            </intent-filter>
        </activity>
        <receiver android:name=".BootReceiver">
            <intent-filter>
                <actionandroid:name="android.intent.action.BOOT_COMPLETED"/>
            </intent-filter>
        </receiver>
    </application>
</manifest>
```

（6）运行项目，测试运行效果。

通常，在广播接收器的 OnReceive()方法中完成广播消息处理逻辑。OnReceive()方法中不允许开启线程，所以最好不要在启动时添加过多的逻辑和耗时的操作。OnReceive()方法太长时，程序就会报错，所以，一般广播接收器都用于完成创建通知、启动服务、启动应用等类似的各种操作。

4.5 小结

本章主要介绍了 Android 的广播机制。广播机制与 Activity 之间使用 Intent 传递数据类似，Android 将广播消息封装在 Intent 对象中并传递给接收器。广播消息可以同时发送给多个接收器（系统广播就发送给系统中的全体应用）。接收器中通过 IntentFilter 来设置可接收的广播，只有与 IntentFilter 配置一致的广播消息才能被接收。

接收器是一个 BroadcastReceiver 类的子类，在其 OnReceive()方法中完成广播消息的处理。接收器只能在注册后才能接收广播，可在程序清单文件 AndroidManifest.xml 中进行静态注册，也可在代码中动态注册和注销。

4.6 习题

1. Android 中的广播可分哪些类型？分别有什么特点？
2. 简述一个广播接收器类的基本结构。
3. 简述如何注册广播接收器。
4. 简述如何使用有序广播。

PART05

第5章

数据存储

重点知识：

- 文件存储
- 共享存储
- SQLite数据库存储

■ 当我们在使用 QQ、微信或者其他应用程序时，都是在不停地和各种类型的数据打交道。数据可以说是应用程序的核心，所以数据存储就成了应用程序不可或缺的基石。

Android 系统主要提供了 3 种数据存储方式，包括文件存储、共享存储和 SQLite 数据库存储。本章将介绍这 3 种数据存储方式。

5.1 文件存储

文件是一种基本的数据存储方式，适合于存储简单的文本或二进制数据。在使用文件时，可将其存放在内部存储器或外部存储器（如 SD 卡等）中。

5.1.1 读写内部存储文件

默认情况下，保存到内部存储器中的文件是当前应用的私有文件，其他应用或用户不能访问。在卸载应用时，文件也会随之删除。

读写内部存储的
文件数据

Context 类的 openFileOutput()方法用于打开一个内部存储文件并向文件写入数据，其基本格式如下。

```
FileOutputStream fos = openFileOutput(FILENAME, Context.MODE_PRIVATE);
```

openFileOutput()方法的第 1 个参数为文件名。需注意的是，文件名中不能包含路径。第 2 个参数为访问模式：MODE_PRIVATE 为默认模式，表示当指定文件存在时，原来的文件会被覆盖；MODE_APPEND 表示当指定文件存在时，写入的数据会添加到文件末尾。较早版本的 Android 还提供另外两种文件访问模式，为 MODE_WORLD_READABLE 和 MODE_WORLD_WRITEABLE，因为这两种模式容易引起安全漏洞，所以在 Android 4.2 版本中已被废弃。

openFileOutput()方法会返回一个 FileOutputStream 对象。使用该对象可将数据写入文件。例如，如下代码即可将一个字符串写入内部存储文件。实例项目：源代码\05\UseInternalStorage。

```
String FILENAME = "myfile";
String data = "在内部文件中的数据";
try {
    FileOutputStream fos = openFileOutput(FILENAME, Context.MODE_PRIVATE);
    OutputStreamWriterosw=new OutputStreamWriter(fos);
    osw.write(data);
    osw.flush();
    fos.flush();
    osw.close();
    fos.close();
} catch (Exception e) {
    e.printStackTrace();
}
```

类似的，Context 类还提供了一个 openFileInput()方法，用于打开一个内部存储文件并从文件读取数据。openFileInput()方法可以返回一个 FileInputStream 对象，使用该对象可从文件读取数据。

例如，如下代码即可读出文件中的字符串。实例项目：源代码\05\UseInternalStorage。

```
try {
    FileInputStream fis = openFileInput(FILENAME);
    InputStreamReader isr=new InputStreamReader(fis,"UTF-8");
    char[] data2=new char[fis.available()];
    isr.read(data2);
    isr.close();
```

```
        fis.close();
        TextViewtextView=(TextView)findViewById(R.id.textView);
        textView.setText(new String(data2));
} catch (Exception e) {
        e.printStackTrace();
}
```

5.1.2 读写外部存储文件

读取外部存储的文件数据

内部存储使用设备自带的内部存储空间，外部存储使用设备出厂时不存在、用户使用时添加的外部存储介质，例如 TF 卡、SD 卡等。

要访问外部存储中的文件，首先应用必须具有 READ_EXTERNAL_STORAGE（读）或 WRITE_EXTERNAL_STORAGE（写）权限（写权限包含了读权限）。可在应用的清单文件 AndroidManifest.xml 中为应用申请权限，例如如下代码。

```
<manifest …>
    <uses-permission android:name="android.permission.WRITE_EXTERNAL_STORAGE"/>
    …
</manifest>
```

即使为应用申请了权限，在应用安装到设备中之后，还需要在设备的“设置/应用管理”中找到该应用并为其启用存储访问权限，否则将仍然无法访问外部存储。

内置的外部存储路径通常是/storage/emulated/0 或者/mnt/sdcard。不同设备可能有所区别。用下面的方法可获得外部存储路径。

```
Filesdcard=Environment.getExternalStorageDirectory();
```

在使用外部存储之前，还应检测其状态是否可用，例如如下代码。

```
private boolean isReadable(){//检测存储卡是否可读
    String state = Environment.getExternalStorageState();
    if (state.equals(Environment.MEDIA_MOUNTED) ||
            state.equals(Environment.MEDIA_MOUNTED_READ_ONLY)) {
        return true;
    }
    returnfalse;
}
private boolean isWriteable(){//检测存储卡是否可写
    String state = Environment.getExternalStorageState();
    if (state.equals(Environment.MEDIA_MOUNTED)) {
        return true;
    }
    returnfalse;
}
```

如下代码实现了将数据写入外部存储文件以及从外部存储文件读取数据。实例项目：源代码\05\UseExternalStorage。

```
Button btn1=(Button)findViewById(R.id.button1) ;
btn1.setOnClickListener(new View.OnClickListener() {
    @Override
    public void onClick(View v) {
        if(!isWriteable()){
            textView.setText("SD卡不可用");
            return;
        }
        File sdcard=Environment.getExternalStorageDirectory();
        File mf=new File(sdcard,"myfile.txt");
        //File mf=new File("/storage/emulated/0/myfile.txt");
        try {
            mf.createNewFile();
            String data = "在SD卡文件中的数据";
            FileOutputStream fos = new FileOutputStream(mf);
            OutputStreamWriter osw=new OutputStreamWriter(fos);
            osw.write(data);
            osw.flush();
            fos.flush();
            osw.close();
            fos.close();
            textView.setText("数据已经写入外部文件");
        } catch (Exception e) {
            textView.setText(e.getMessage());
        }
    }
});
Button btn2=(Button)findViewById(R.id.button2) ;
btn2.setOnClickListener(new View.OnClickListener() {
    @Override
    public void onClick(View v) {
        if(!isReadable()){
            textView.setText("SD卡不可读");
            return;
        }
        File sdcard=Environment.getExternalStorageDirectory();
        File mf=new File(sdcard,"myfile.txt");
        try {
            FileInputStream fis = new FileInputStream(mf);
            InputStreamReader isr=new InputStreamReader(fis,"UTF-8");
            char[] data2=new char[fis.available()];
            isr.read(data2);
            isr.close();
```

```
                fis.close();
                textView.setText(new String(data2));
            } catch (Exception e) {
                textView.setText(e.getMessage());
            }
        }
});
```

5.1.3 应用的私有文件

Environment.getExternalStorageDirectory()可以返回第 1 个外部存储根目录。目前，绝大多数设备的第 1 个外部存储已经内置到设备中，建议不要直接访问外部存储卡的根目录。如果存储数据的文件仅仅在当前应用中使用，则可以使用应用程序私有的外部存储路径。调用 Context.getExternalFilesDir()方法可获得应用程序的私有外部存储路径。存储在私有外部存储路径中的文件称为应用的私有文件，这些文件会随着应用程序的卸载被删除。

例如，如下代码即实现了创建私有文件。

```
Fileprivatepath=Context.getExternalFilesDir();
File mf=new File(privatepath,"myfile.txt");
```

5.1.4 访问公共目录

Android 允许应用将文件存放到公共目录中，例如 documents、download、music 等，以便与其他应用分享数据。要获得相应公共目录的 File，可调用 Environment 的 getExternalStoragePublicDirectory()方法，其参数为目录类型，例如 DIRECTORY_MUSIC、DIRECTORY_PICTURES 或其他类型。

例如，如下代码可在公共目录 documents 中创建一个 TXT 文件。实例项目：源代码\05\UsePublicPath。

```
If(!isWritable()){
   Toast.makeText(this,"SD卡不可用",Toast.LENGTH_SHORT).show();
   return;
}
Filesdcard=
            Environment.getExternalStoragePublicDirectory(Environment.DIRECTORY_DOCUMENTS);
File mf=new File(sdcard,"myfile.txt");
try {
    mf.createNewFile();
    Toast.makeText(this,"成功创建文件",Toast.LENGTH_SHORT).show();
} catch (Exception e) {
    Toast.makeText(this,e.getMessage(),Toast.LENGTH_SHORT).show();
}
```

5.2 共享存储

共享存储采用 SharedPreferences 文件来保存数据。SharedPreferences 虽然也用文件来保存数据，但它使用键值对的方式存储数据。在保存数据时，需要为数据提供一个相对唯一的键；在读取数据时，通过键把相应的值取出。SharedPreferences 支持多种不同类型的数据存储，包括 boolean、int、long、float、String 以及 Set<String>等。

Android 本地数据存储之 Shared Preferences

因为 SharedPreferences 文件使用键值对的方式存取数据，所以在存取大量数据时效率不是很高，更适用于存储应用的个性化参数设置。

5.2.1 将数据存入 SharedPreferences 文件

要将数据存入 SharedPreferences 文件，需要下列几个步骤。

（1）获得 SharedPreferences 对象。

（2）获得 SharedPreferences 对象的 Editor 对象。

（3）调用 Editor 对象的方法，向文件添加数据。

（4）提交数据，完成数据存储操作。

1. 获得 SharedPreferences 对象

Android 提供了 3 种获得 SharedPreferences 对象的方法。

第 1 种获得 SharedPreferences 对象的方法是调用 Context 类的 getSharedPreferences()方法，例如如下代码。

```
SharedPreferences pref=getSharedPreferences("myPreferences",MODE_PRIVATE);
```

getSharedPreferences()方法的第 1 个参数是 SharedPreferences 文件的名称，第 2 个参数是操作模式。MODE_PRIVATE 是默认的操作模式，等价于 0。MODE_PRIVATE 表示文件属于当前应用的私有文件，其他应用不能访问。另一种模式是 MODE_APPEND，表示当指定文件存在时向文件中添加数据。

第 2 种获得 SharedPreferences 对象的方法是调用 Activity 类的 getPreferences()方法，例如如下代码。

```
SharedPreferences pref=getPreferences(MODE_PRIVATE);
```

getPreferences()方法的参数指定文件操作模式，它默认以当前活动的类名作为 SharedPreferences 文件的名称。

第 3 种获得 SharedPreferences 对象的方法是调用 PreferenceManager 类的 getDefault SharedPreferences()方法，例如如下代码。

```
SharedPreferences pref= PreferenceManager.getDefaultSharedPreferences(this);
```

getDefaultSharedPreferences()方法的参数为当前应用上下文，它默认以当前应用的包名作为 SharedPreferences 文件的名称。

2. 获得 SharedPreferences 对象的 Editor 对象

调用 SharedPreferences 对象的 edit()方法可创建一个 Editor 对象，例如如下代码。

```
SharedPreferences.Editoreditor=pref.edit();
```

3. 调用 Editor 对象的方法，向文件添加数据

调用 Editor 对象的各种 putXXX()方法，都可向 SharedPreferences 文件添加数据，例如如下代码。

```
editor.putString("username","admin");
editor.putString("password","12345");
```

```
editor.putBoolean("remembered",true);
```

putXXX()方法的第 1 个参数为键，第 2 个参数为通过键保存的数据（值）。

4．提交数据，完成数据存储操作

在调用 putXXX()方法添加了数据后，必须调用 Editor 对象的 apply()方法提交数据，才能将数据存入文件，最终完成数据存储操作，例如如下代码。

```
editor.apply();
```

5.2.2 读取 SharedPreferences 文件数据

获得 SharedPreferences 对象后，调用相应的 getXXX()方法即可读取存储在文件中的数据，例如如下代码。

```
booleanisRemembered=pref.getBoolean("remembered",false);
etName.setText(pref.getString("username",""));
etPwd.setText(pref.getString("password",""));
```

getXXX()方法的第 1 个参数为键，第 2 个参数为默认值。如果 SharedPreferences 文件中无指定的键，则 getXXX()方法会返回第 2 个参数指定的默认值。

5.2.3 实现记住密码功能

通常，应用的登录界面中往往会提供一个记住密码功能，避免用户下一次登录时再次输入登录信息。

这里将使用 SharedPreferences 文件来存储登录界面中输入的登录信息，具体操作步骤如下。实例项目：源代码\05\UseSharedPreferences。

（1）在 AndroidStudio 中创建一个新项目，将应用名称设置为 UseSharedPreferences，并为项目添加一个空活动。

（2）修改 activity_main.xml，在主活动布局中添加控件，代码如下。

```
<?xml version="1.0" encoding="utf-8"?>
<LinearLayout xmlns:android="http://schemas.android.com/apk/res/android"
    xmlns:tools="http://schemas.android.com/tools"
    android:id="@+id/activity_main"  android:layout_width="match_parent"
    android:layout_height="match_parent" android:orientation="vertical"
    tools:context="com.example.xbg.usesharedpreference.MainActivity">
    <LinearLayout
        android:orientation="horizontal" android:layout_width="match_parent"
        android:layout_height="wrap_content">
        <TextView
            android:layout_height="wrap_content" android:layout_width="wrap_content"
            android:text="用户名：" android:textSize="20sp"
            android:layout_gravity="right|center_vertical" />
        <EditText
            android:id="@+id/etName"
            android:layout_weight="1"
            android:layout_width="wrap_content" android:layout_height="wrap_content" />
    </LinearLayout>
```

```
    <LinearLayout
        android:orientation="horizontal"
        android:layout_width="match_parent"
        android:layout_height="wrap_content">
        <TextView
            android:layout_width="wrap_content"
            android:layout_height="wrap_content"
            android:text="密    码："
            android:textSize="20sp"
            android:layout_gravity="right|center_vertical" />
        <EditText
            android:id="@+id/etPwd"
            android:layout_weight="1"
            android:inputType="textPassword"
            android:layout_width="wrap_content"
            android:layout_height="wrap_content" />
    </LinearLayout>
    <CheckBox
        android:text="记住密码"
        android:layout_width="match_parent"
        android:layout_height="wrap_content"
        android:id="@+id/cbRemember" />
    <Button
        android:text="登录"
        android:layout_width="match_parent"
        android:layout_height="wrap_content"
        android:id="@+id/btLogin" />
</LinearLayout>
```

（3）修改 MainActivity.java。在 onCreate()方法中读取 SharedPreferences 文件中保存的登录信息，初始化登录界面。同时，添加按钮的单击事件监听器。在单击按钮时，如果登录信息正确，且用户选择记住密码，则将当前登录信息存入文件，代码如下。

```
package com.example.xbg.usesharedpreference;
import android.content.SharedPreferences;
...
public class MainActivity extends AppCompatActivity {
    private   EditText etName;
    private   EditText etPwd;
    private CheckBox cbRemember;
    private   Button btLogin;
    private SharedPreferences pref;
    private SharedPreferences.Editor editor;
    @Override
    protected void onCreate(Bundle savedInstanceState) {
        super.onCreate(savedInstanceState);
```

```
        setContentView(R.layout.activity_main);
        pref= PreferenceManager.getDefaultSharedPreferences(this);
        boolean isRemembered=pref.getBoolean("remembered",false);
        etName= (EditText) findViewById(R.id.etName);
        etPwd= (EditText) findViewById(R.id.etPwd);
        cbRemember= (CheckBox) findViewById(R.id.cbRemember);
        if(isRemembered){//根据存储的数据初始化界面
            etName.setText(pref.getString("username",""));
            etPwd.setText(pref.getString("password",""));
            cbRemember.setChecked(true);
        }
        btLogin= (Button) findViewById(R.id.btLogin);
        btLogin.setOnClickListener(new View.OnClickListener() {
            @Override
            public void onClick(View v) {
                if(CheckLog()){
                    //登录信息正确，判断是否需要保存当前登录信息
                    if(cbRemember.isChecked()){
                        //保存登录信息
                        String id=etName.getText().toString();
                        String pwd=etPwd.getText().toString();
                        editor=pref.edit();
                        editor.putString("username",id);
                        editor.putString("password",pwd);
                        editor.putBoolean("remembered",true);
                        Toast.makeText(MainActivity.this,"登录信息已保存！",
                                Toast.LENGTH_SHORT).show();
                    }else{
                        editor.clear();//清除SharedPreferences数据
                    }
                    editor.apply();//使SharedPreferences修改生效
                    Toast.makeText(MainActivity.this,"登录成功！",
                                            Toast.LENGTH_SHORT).show();
                }else{
                    Toast.makeText(MainActivity.this,"用户名或密码错误！",
                                            Toast.LENGTH_SHORT).show();
                }
            }
        });
    }
    private   boolean CheckLog(){
        //判断用户登录信息是否正确
        String id=etName.getText().toString();
        String pwd=etPwd.getText().toString();
        //如果用户名是admin，密码是123，则认为登录信息正确
```

```
            if(id.equals("admin") && pwd.equals("123")){
                return   true;
            }else{
                return false;
            }
        }
}
```

（4）运行项目，测试运行效果，如图 5-1 所示。

图 5-1　记住密码

5.3　SQLite 数据库存储

Android 本地数据存储之 SQLite

SQLite 是一款轻量级的关系数据库。它运算速度快，运行内存少，只需几百 KB 的内存即可，因而适用于移动设备。SQLite 不仅支持标准的 SQL 语法，还支持 ACID 失误。Android 系统内置了 SQLite 数据库，使得在 Android 应用中可以轻松使用数据库来完成数据存储。

5.3.1　创建数据库

Android 提供了一个抽象类 SQLiteOpenHelper。借助该类，可以很方便地实现数据库的创建、升级以及数据的管理。

SQLiteOpenHelper 是一个抽象类，所以需要创建一个类来继承它，并实现需要的方法。SQLiteOpenHelper 的子类必须实现 onCreate()和 OnUpgrade()两个方法。onCreate()方法在创建数据库时被调用，完成数据库的初始化操作，例如创建数据表、添加初始数据等。OnUpgrade()方法在升级数据库时被调用。

SQLiteOpenHelper 类提供了 getWritableDatabase()和 getReadableDatabase()两个方法用于打开或创建数据库。如果指定的数据库存在，则直接打开，否则创建一个新的数据库。getWritableDatabase()和 getReadableDatabase()都返回一个 SQLiteDatabase 实例对象，通过该对象完成对数据库的各种操作。如果数据库无法写入数据（如磁盘空间已满），getReadableDatabase()将返回一个只读数据库对象，此时使用 getWritableDatabase()方法则会出错。

SQLiteOpenHelper 类提供了如下两个构造方法。

```
SQLiteOpenHelper(Context context, String name, SQLiteDatabase.CursorFactory factory, int version)
SQLiteOpenHelper(Context context, String name, SQLiteDatabase.CursorFactory factory, int version, DatabaseErrorHandler errorHandler)
```

参数 context 为上下文对象，name 为数据库名称，factory 是用于创建保存查询结果的自定义 Cursor 对象（一般使用 null 表示使用默认的 Cursor 对象），version 为数据库版本号（从 1 开始）。

如下代码中的 MySQLiteHelper 类即是 SQLiteOpenHelper 的子类。实例项目：源代码\05\UseSQLite。

```
package com.example.xbg.usesqlite;
import android.content.Context;
import android.database.sqlite.SQLiteDatabase;
import android.database.sqlite.SQLiteOpenHelper;
import android.widget.Toast;
public class MySQLiteHelper extends SQLiteOpenHelper {
    private static String CREATE_TABLE_USER="create table users("+
              "id integer primary key autoincrement,"+
              "userid text,password text)";
    private Context sContext;
    public MySQLiteHelper(Context context, String name,
                          SQLiteDatabase.CursorFactory factory, int version) {
        super(context, name, factory, version);
        sContext=context;
    }
    @Override
    public void onCreate(SQLiteDatabase db) {
        //执行数据库初始化操作
        db.execSQL(CREATE_TABLE_USER);
        Toast.makeText(sContext," 成功创建数据表",Toast.LENGTH_LONG).show();
    }
    @Override
    public void onUpgrade(SQLiteDatabase db, int oldVersion, int newVersion) {
        //执行数据库升级操作
    }
}
```

在调用 getWritableDatabase()和 getReadableDatabase()方法时，如果指定的数据库不存在，首先会创建，onCreate()方法的参数为引用新建数据库的 SQLiteDatabase 对象。如果数据库已经存在，则不会调用 onCreate()方法。在本例的 onCreate()方法中，通过 SQLiteDatabase 对象调用 execSQL()方法执行 SQL 命令，为数据库创建了一个数据表 users，但这不是必需的，如果 onCreate()方法没有为数据库添加数据表，则数据库为空数据库。

SQLiteDatabase 对象的 execSQL()方法可用于执行非 Select 查询和其他不返回结果的 SQL 命令。

在 MainActivity 中创建 MySQLiteHelper 类对象，即可用于创建数据库，代码如下。

```
public class MainActivity extends AppCompatActivity {
    privateMySQLiteHelpersqLiteHelper;
    privateSQLiteDatabasemyDb;
    TextViewtvPath;
    @Override
    protected void onCreate(Bundle savedInstanceState) {
        super.onCreate(savedInstanceState);
        setContentView(R.layout.activity_main);
        tvPath= (TextView) findViewById(R.id.txtPath);
        Button btCreateDb=(Button)findViewById(R.id.btCreateDb);
        btCreateDb.setOnClickListener(new View.OnClickListener() {
            @Override
            public void onClick(View v) {
                sqLiteHelper=new MySQLiteHelper(MainActivity.this,"usersdb.db",null,1);
                myDb=sqLiteHelper.getWritableDatabase();//完成创建数据库
                String path=myDb.getPath();
                tvPath.setText("数据库："+path);//显示数据库文件及其路径
            }
        });
...
    }
}
```

利用 SQLiteOpenHelper 类创建的 SQLite 数据库的文件默认在/data/user/0/包名/databases/目录中，不同设备可能有所区别。

运行上述程序，执行创建数据库操作，效果如图 5-2 所示。成功创建数据库后，文本视图控件中会显示完整的数据库文件名，屏幕下方会显示 Toast 信息，提示成功创建数据表。

要删除 SQLite 数据库，可调用 SQLiteOpenHelper 类的实例方法 deleteDatabase()，代码如下。

```
Button btDeleteDb=(Button)findViewById(R.id.btDeleteDb);
btDeleteDb.setOnClickListener(new View.OnClickListener() {
    @Override
    public void onClick(View v) {
    if(myDb.isOpen()){
        myDb.close();                          //若数据库已打开，则先将其关闭
    }
```

```
        String path=myDb.getPath();              //获得数据库文件名（含路径）
        File db=new File(path);
        SQLiteDatabase.deleteDatabase(db);       //删除数据库
        tvPath.setText("数据库已删除");             //显示数据库文件及其路径
    }
});
```

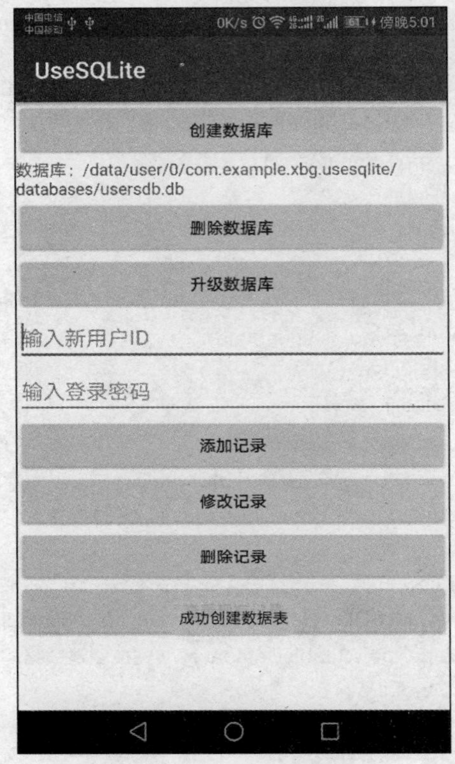

图 5-2　创建数据库后的效果

5.3.2　升级数据库

SQLiteOpenHelper 类的构造方法中的数据库版本号参数用于升级或降级数据库。若提供的版本号比当前版本号大，则调用 onUpgrade()方法升级当前数据库；如果提供的版本号比当前版本号小，则调用 onDowngrade()方法对数据库进行降级。

5.3.1 小节中创建了一个数据库 usersdb.db，并为其创建了一个 users 表。users 表用于保存用户 ID 和登录密码。下面增加一个数据表保存用户类型，且不同类型具有不同的权限，同时实现升级数据库功能，且在升级时重建数据库中的表，可进行如下操作。

（1）修改 MySQLiteHelper 类，代码如下。实例项目：源代码\05\UseSQLite。

```
package com.example.xbg.usesqlite;
import android.content.Context;
...
public class MySQLiteHelper extends SQLiteOpenHelper {
    private static String CREATE_TABLE_USER="create table users("+
            "id integer primary key autoincrement,"+
```

```
                "userid text,password text)";
    private static String CREATE_TABLE_TYPE="create table types("+
                "id integer primary key autoincrement,"+
                "type_code,describe text)";
…
    public void onCreate(SQLiteDatabase db) {
        //执行数据库初始化操作
        db.execSQL(CREATE_TABLE_USER);
        db.execSQL(CREATE_TABLE_TYPE);
        Toast.makeText(sContext,"成功创建数据表",Toast.LENGTH_LONG).show();
    }
    @Override
    public void onUpgrade(SQLiteDatabase db, int oldVersion, int newVersion) {
        //执行数据库升级操作
        db.execSQL("drop table if exists users");
        db.execSQL("drop table if exists types");
        onCreate(db);
    }
}
```

（2）修改 MainActivity，添加一个按钮来执行数据库升级操作，代码如下。

```
Button btUpgradeDb=(Button)findViewById(R.id.btUpgradeDb);
btUpgradeDb.setOnClickListener(new View.OnClickListener() {
    @Override
    public void onClick(View v) {
    sqLiteHelper=new MySQLiteHelper(MainActivity.this,"usersdb.db",null,2);
    myDb=sqLiteHelper.getWritableDatabase();//完成数据库升级
    }
});
```

5.3.3 添加数据

SQLiteDatabase 对象的 insert()方法用于为表添加数据。insert()方法的基本格式如下。

```
insert(String table, String nullColumnHack, ContentValues values)
```

参数 table 指定要添加记录的表的名称。参数 nullColumnHack 指定记录中需要赋值为 null 的列名，用 null 作为参数可表示没有列需要赋值为 null。参数 values 包含要添加的数据。

如下代码实现了在单击按钮时将用户输入的数据添加到表中。实例项目：源代码\05\UseSQLite。

```
Button btAdd=(Button)findViewById(R.id.btAdd);
btAdd.setOnClickListener(new View.OnClickListener() {
    @Override
    public void onClick(View v) {
        if(myDb==null){
            return;         //在没有创建数据库时，不执行添加数据操作
        }
        ContentValues cv=new ContentValues();
        EditTextetID= (EditText) findViewById(R.id.etName);
```

```
        EditTextetPwd= (EditText) findViewById(R.id.etPassword);
        String name=etID.getText().toString();
        String password=etPwd.getText().toString();
        cv.put("userid",name);
        cv.put("password",password);
        myDb.insert("users",null,cv);                    //将数据添加到数据表
        Toast.makeText(MainActivity.this,"成功添加记录",Toast.LENGTH_LONG).show();
        refreshList();
    }
});
```

5.3.4 更新数据

SQLiteDatabase 对象的 update()方法用于更新数据。update ()方法的基本格式如下。

```
update(String table, ContentValues values, String whereClause, String[] whereArgs)
```

参数 table 指定要更新的数据所在的表的名称。参数 values 包含要更新的列的值。参数 whereClause 指定记录筛选条件，只有符合条件的记录才修改指定列，其中用问号指定需要填充的参数。whereArgs 指定要填充到 whereClause 中的参数值。update()方法会返回被更新的记录数。

如下代码实现了在单击按钮时利用用户输入的数据更新表中的记录。实例项目：源代码\05\UseSQLite。

```
Button btUpdate=(Button)findViewById(R.id.btUpdate);
btUpdate.setOnClickListener(new View.OnClickListener() {
    @Override
    public void onClick(View v) {
        if(myDb==null){
            return;//在没有创建数据库时，不执行后续操作
        }
        EditTextetID= (EditText) findViewById(R.id.etName);
        EditTextetPwd= (EditText) findViewById(R.id.etPassword);
        String name=etID.getText().toString();
        String password=etPwd.getText().toString();
        ContentValues cv=new ContentValues();
        cv.put("password",password);
        myDb.update("users",cv,"userid=?",new String[]{name});
        Toast.makeText(MainActivity.this,"成功修改记录",Toast.LENGTH_LONG).show();
        refreshList();
    }
});
```

5.3.5 删除数据

SQLiteDatabase 对象的 delete()方法用于删除数据。delete()方法的基本格式如下。

```
delete(String table, String whereClause, String[] whereArgs)
```

各参数意义与 update()方法类似，且只有符合条件的记录才被删除。delete()方法会返回被删除的记录数。

如下代码实现了在单击按钮时将用户输入的数据作为条件删除表中的记录。实例项目：源代码\05\UseSQLite。

```
Button btDelete=(Button)findViewById(R.id.btDelete);
btDelete.setOnClickListener(new View.OnClickListener() {
    @Override
    public void onClick(View v) {
        if(myDb==null){
            return;//在没有创建数据库时，不执行后续操作
        }
        EditTextetID= (EditText) findViewById(R.id.etName);
        String name=etID.getText().toString();
        myDb.delete("users","userid=?",new String[]{name});
        Toast.makeText(MainActivity.this,"成功删除记录",Toast.LENGTH_LONG).show();
        refreshList();
    }
});
```

5.3.6 查询数据

SQLiteDatabase 对象的 query()方法用于查询数据。常用的 query()方法有下列 3 种基本格式。

```
query(boolean distinct, String table, String[] columns, String selection, String[] selectionArgs, String groupBy, String having, String orderBy, String limit)
query(String table, String[] columns, String selection, String[] selectionArgs, String groupBy, String having, String orderBy, String limit)
query(String table, String[] columns, String selection, String[] selectionArgs, String groupBy, String having, String orderBy)
```

各个参数的含义如下所述。

- distinct：为 true 时，表示返回结果中不包含重复值。
- table：指定查询的表名称，对应 SQL SELECT 命令的“from 表名称”部分。
- columns：指定查询结果中包含的列名称，对应 SQL SELECT 命令的“select 列名称 1,列名称 2,…”部分。
- selection：指定记录筛选条件，对应 SQL SELECT 命令的“where 条件”部分。
- selectionArgs：指定填充筛选条件占位符的参数。
- groupBy：指定查询分组列名称，对应 SQL SELECT 命令的“group by 列名称 1,列名称 2,…”部分。
- having：指定查询分组的条件，对应 SQL SELECT 命令的“having 条件”部分。
- orderBy：指定查询结果排序列名称，对应 SQL SELECT 命令的“order by 列名称 1,列名称 2,…”部分。
- limit：指定返回的查询结果的最大记录数。

query()方法会返回一个包含查询结果的 Cursor 对象，通过该对象可逐条读取查询结果中的记录。

如下代码实现了在单击按钮时查询 users 表中的数据，并使用 Toast 显示。实例项目：源代码\05\UseSQLite。

```
Button btGetAll=(Button)findViewById(R.id.btGetAll);
btGetAll.setOnClickListener(new View.OnClickListener() {
    @Override
```

```
    public void onClick(View v) {
        if(myDb==null){
            return;//在没有创建数据库时，不执行后续操作
        }
        Cursor c=myDb.query("users",null,null,null,null,null,null);
        String msg="当前记录如下：\n";
        if(c.moveToFirst()){
            do{
                msg=msg+"userid:"+c.getString(c.getColumnIndex("userid"))+
                "   password="+c.getString(c.getColumnIndex("password"))+"\n";
            }while(c.moveToNext());
        }
        Toast.makeText(MainActivity.this,msg,Toast.LENGTH_LONG).show();
    }
});
```

也可使用 Cursor 对象创建适配器来填充 ListView 控件。在填充 ListView 控件时，首先可自定义一个列表项的布局，该布局定义每条记录中的各个列如何显示，例如如下代码。

```
<?xml version="1.0" encoding="utf-8"?>
<LinearLayoutxmlns:android="http://schemas.android.com/apk/res/android"
    android:orientation="horizontal" android:layout_width="match_parent"
    android:layout_height="match_parent">
    <TextView
        android:text="用户ID："
        android:layout_width="wrap_content"android:layout_height="wrap_content" />
    <TextView android:layout_width="wrap_content"android:layout_height="wrap_content"
        android:id="@+id/textID"    android:layout_weight="1" />
    <TextView android:text="密码："
        android:layout_width="wrap_content" android:layout_height="wrap_content"/>
    <TextView android:layout_width="wrap_content"   android:layout_height="wrap_content"
        android:id="@+id/textPwd"   android:layout_weight="1" />
</LinearLayout>
```

该布局水平显示一条记录中的用户 ID 和密码信息。

在 Activity 中，用 Cursor 对象创建 SimpleCursorAdapter 对象来填充 ListView 控件的代码如下。实例项目：源代码\05\UseSQLite。

```
private void refreshList(){
    Cursor c=myDb.query("users",new String[]{"id as _id","userid","password"},
            null,null,null,null,null);
    SimpleCursorAdapter adapter=new SimpleCursorAdapter(MainActivity.this,
            R.layout.recordlist, c,new String[]{"userid","password"},
            new int[]{R.id.textID,R.id.textPwd});
    ListView lv= (ListView) findViewById(R.id.lvRecords);
    lv.setAdapter(adapter);
}
```

图 5-3 所示为上述程序的运行效果，在 Toast 信息和界面最下方的列表中显示了数据库的 users 表数据。

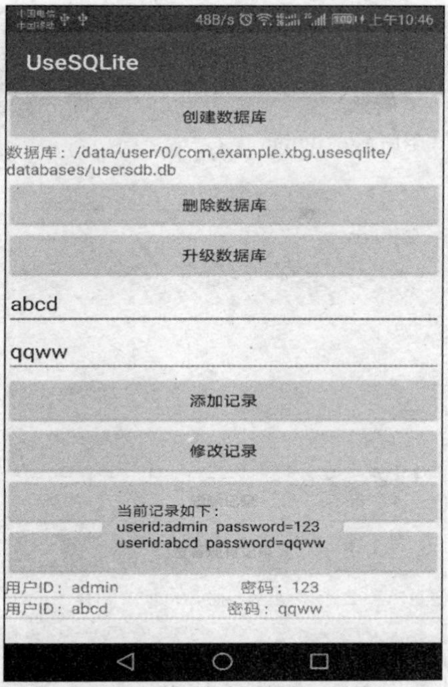

图 5-3　不同方式下显示的数据

创建 SimpleCursorAdapter 对象时，Cursor 中的记录必须包含一个“_id”字段，否则程序会出错。

在 UseSQLite 实例项目中进行添加、修改、删除操作时，均须调用 refreshList()方法来更新 ListView 控件，使用户看到操作后表中的最新记录数据。

5.3.7　执行 SQL 命令操作数据库

前文利用 SQLiteDatabase 类提供的方法很方便地完成了数据的添加（Insert）、修改（Update）、删除（Delete）和查询（Select）等操作。SQLiteDatabase 类也支持 SQL 命令来完成各种数据操作，简单介绍如下。

1. 添加记录

```
myDb.execSQL("insert into users(userid,password) values(?,?)", new String[]{name,password});
```

2. 修改记录

```
myDb.execSQL("update users set password=? where userid=?",new String[]{password,name});
```

3. 删除记录

```
myDb.execSQL("delete from users where userid=?",new String[]{name});
```

4. 查询记录

```
Cursor c=myDb.rawQuery("select id as _id,userid,password from users",null);
```

5.4　编程实践：基于数据库的登录验证

本节将综合应用本章所学知识，实现基于数据库的登录验证，程序运行效果如图 5-4 所示。

图 5-4　登录验证

具体操作步骤如下。

（1）在 AndroidStudio 中创建一个新项目，将应用名称设置为 LoginFromDatabase，并为项目添加一个空活动。

（2）修改 activity_main.xml，为主活动布局添加一个按钮控件，代码如下。

```
<?xml version="1.0" encoding="utf-8"?>
<LinearLayoutxmlns:android="http://schemas.android.com/apk/res/android"
    xmlns:tools="http://schemas.android.com/tools"
    android:id="@+id/activity_main"
    android:layout_width="match_parent"
    android:layout_height="match_parent"
    android:paddingBottom="@dimen/activity_vertical_margin"
    android:paddingLeft="@dimen/activity_horizontal_margin"
    android:paddingRight="@dimen/activity_horizontal_margin"
    android:paddingTop="@dimen/activity_vertical_margin"
    android:orientation="vertical"
    tools:context="com.example.pad.loginfromdatabase.MainActivity">
    <LinearLayout
        android:orientation="horizontal" android:layout_width="match_parent"
        android:layout_height="wrap_content">
        <TextView
            android:layout_height="wrap_content" android:layout_width="wrap_content"
            android:text="用户名：" android:textSize="20sp"
            android:layout_gravity="right|center_vertical" />
```

```
        <EditText
            android:id="@+id/etName"
            android:layout_weight="1"
            android:layout_width="wrap_content" android:layout_height="wrap_content" />
    </LinearLayout>
    <LinearLayout
        android:orientation="horizontal"
        android:layout_width="match_parent"
        android:layout_height="wrap_content">
        <TextView
            android:layout_width="wrap_content"
            android:layout_height="wrap_content"
            android:text="密    码："
            android:textSize="20sp"
            android:layout_gravity="right|center_vertical" />
        <EditText
            android:id="@+id/etPwd"
            android:layout_weight="1"
            android:inputType="textPassword"
            android:layout_width="wrap_content"
            android:layout_height="wrap_content" />
    </LinearLayout>
    <CheckBox
        android:text="记住密码"
        android:layout_width="match_parent"
        android:layout_height="wrap_content"
        android:id="@+id/cbRemember" />
    <Button
        android:text="登录"
        android:layout_width="match_parent"
        android:layout_height="wrap_content"
        android:id="@+id/btLogin" />
</LinearLayout>
```

（3）添加一个 Java 类，继承 SQLiteOpenHelper 类，在创建数据库时创建一个 users 表，并添加两条记录用于登录验证，代码如下。

```
package com.example.pad.loginfromdatabase;
import android.content.Context;
import android.database.sqlite.SQLiteDatabase;
import android.database.sqlite.SQLiteOpenHelper;
import android.widget.Toast;
public class MySQLiteHelper extends SQLiteOpenHelper {
    private static String CREATE_TABLE_USER="create table users("+
            "id integer primary key autoincrement,"+
            "userid text,password text,remembered integer)";
    private Context sContext;
```

```
    publicMySQLiteHelper(Context context, String name, SQLiteDatabase.CursorFactoryfactory, int version) {
        super(context, name, factory, version);
        sContext=context;
    }
    @Override
    public void onCreate(SQLiteDatabase db) {
        db.execSQL(CREATE_TABLE_USER);//创建表
        //添加记录用于登录验证
        db.execSQL("insert into users(userid,password,remembered) values(?,?,0)",
                new String[]{"admin","123456"});
        db.execSQL("insert into users(userid,password,remembered) values(?,?,0)",
                new String[]{"mike","123456"});
        Toast.makeText(sContext,"数据库初始化成功！",Toast.LENGTH_LONG).show();
    }
    @Override
    public void onUpgrade(SQLiteDatabase db, int oldVersion, int newVersion) {
    }
}
```

（4）修改 MainActivity.java，代码如下。

```
package com.example.pad.loginfromdatabase;
import android.content.SharedPreferences;
^
public class MainActivity extends AppCompatActivity {
    privateEditTextetName;
    privateEditTextetPwd;
    private CheckBox cbRemember;
    private Button btLogin;
    privateSharedPreferences pref;
    privateSharedPreferences.Editor editor;
    privateMySQLiteHelpersqLiteHelper;
    privateSQLiteDatabasemyDb;
    @Override
    protected void onCreate(Bundle savedInstanceState) {
        super.onCreate(savedInstanceState);
        setContentView(R.layout.activity_main);
        etName= (EditText) findViewById(R.id.etName);
        etPwd= (EditText) findViewById(R.id.etPwd);
        cbRemember= (CheckBox) findViewById(R.id.cbRemember);

        sqLiteHelper=new MySQLiteHelper(MainActivity.this,"usersdb.db",null,1);
        myDb=sqLiteHelper.getWritableDatabase();//完成创建或打开数据库
        //判断是否记住密码来初始化登录界面
        pref= PreferenceManager.getDefaultSharedPreferences(this);
        editor=pref.edit();
        booleanisRemembered=pref.getBoolean("remembered",false);
```

```
        if(isRemembered){
            //执行查询获得当前记住密码的用户登录信息
            Cursor c=myDb.rawQuery("select * from users where remembered=1",null);
            if(c.moveToFirst()){
                //用查询结果初始化界面
                etName.setText(c.getString(c.getColumnIndex("userid")));
                etPwd.setText(c.getString(c.getColumnIndex("password")));
                cbRemember.setChecked(true);
            }
        }
        btLogin= (Button) findViewById(R.id.btLogin);
        btLogin.setOnClickListener(new View.OnClickListener() {
            @Override
            public void onClick(View v) {
                if(CheckLog()){
                    //登录信息正确，判断是否需要保存当前登录信息
                    if(cbRemember.isChecked()){
                        String id=etName.getText().toString();
                        //更新数据库中的记住密码标志
                        myDb.execSQL("update users set remembered=1 where userid=?",
                                          new String[]{id});
                        myDb.execSQL("update users set remembered=0 where userid<>?",
                                          new String[]{id});
                        editor.putBoolean("remembered",true);
                    }else{
                        //更新数据库中的记住密码标志
                        myDb.execSQL("update users set remembered=0");
                        editor.clear();//清除SharedPreferences数据
                    }
                    editor.apply();//使SharedPreferences修改生效
                    Toast.makeText(MainActivity.this,"登录成功！",Toast.LENGTH_SHORT).show();
                }else{
                    Toast.makeText(MainActivity.this,"用户名或密码错误！",
                                                    Toast.LENGTH_SHORT).show();
                }
            }
        });
    }
    private   booleanCheckLog(){
        //判断用户登录信息是否正确
        String id=etName.getText().toString();
        String pwd=etPwd.getText().toString();
        //Cursor c=myDb.rawQuery("select * from users where userid=? and password=?",
                                    new String[]{id,pwd});
        Cursor c=myDb.query("users",new String[]{"userid"},
                "userid=? and password=?",new String[]{id,pwd},null,null,null);
```

```
            if(c.moveToFirst()){
                return   true;
            }else{
                returnfalse;
            }
        }
}
```

（5）运行项目，测试运行效果。

5.5　小结

本章主要介绍了 Android 系统的 3 种主要数据存储方式，包括文件存储、共享存储和数据库存储。文件存储通常在文件中保存字符串等数据，并通过 Java 文件访问相关的类完成数据读写操作。共享存储是利用 SharedPreferences 类提供的相关方法，以键值对的方式读写 SharedPreferences 文件中的数据。数据库存储指利用 Android 内置的 SQLite 数据库来存储各种关系数据。使用 SQLiteOpenHelper 和 SQLiteDatabase 类，可以很方便地完成 SQLite 数据库的各种操作。

5.6　习题

1. 简述如何访问一个内部存储文件。
2. 简述如何访问一个外部存储文件。
3. 简述如何使用 SharedPreferences 共享存储。
4. 简述访问 SQLite 数据库的基本步骤。

PART06

第6章

多媒体

重点知识：

- 播放多媒体文件
- 记录声音
- 使用摄像头和相册

■ 随着社会的发展和技术的不断进步，手机功能已不再仅仅局限于打电话和发短信。Android 提供的多媒体支持功能，使手机的娱乐功能变得更加丰富，例如听音乐、看电影、拍照片等。

本章将介绍如何在 Android 手机中播放多媒体文件、记录声音以及使用摄像头和相册。

6.1 播放多媒体文件

Android 提供了完整的 API 用于播放多媒体文件，开发人员可以轻松实现简易的音频和视频播放 APP。

Android 多媒体之
SoundPool 播放声音

6.1.1 使用 SoundPool 播放音效

SoundPool 类可用于管理和播放应用中的音频资源。这些音频资源既可包含在应用程序中，也可存放于存储器文件中。通常，SoundPool 类只用于播放较短的音频，比如游戏中的各种音效。

要使用 SoundPool 播放音频，首先需创建 SoundPool 对象，代码如下。实例项目：源代码\06\UseSoundPool。

```
if (Build.VERSION.SDK_INT >= 21) {
    SoundPool.Builder builder = new SoundPool.Builder();
    builder.setMaxStreams(2);//设置可加载的音频数量
    //AudioAttributes是一个封装音频各种属性的方法
    AudioAttributes.Builder attrBuilder = new AudioAttributes.Builder();
    attrBuilder.setLegacyStreamType(AudioManager.STREAM_MUSIC);//预设音频类型
    builder.setAudioAttributes(attrBuilder.build());//设置音频类型
    sp = builder.build();//创建SoundPool对象
}
else {//当系统的SDK版本小于21时
    sp = new SoundPool(2, AudioManager.STREAM_SYSTEM, 0);
}
```

在 API 21（Android 5.0）之后的版本中，SoundPool()构造方法已经过时了，需用 SoundPool.Builder 来创建 SoundPool 对象。SoundPool.Builder 对象可执行 setMaxStreams()方法来设置 SoundPool 对象中可加载的最大音频数量。setAudioAttributes()方法则用于设置音频的类型。

SoundPool()构造方法的第 1 个参数为可加载音频的最大数量，第 2 个参数为音频类型，第 3 个参数为声音品质（目前无效，0 为默认）。

获得 SoundPool 对象后，调用 load()方法加载音频资源。load()方法的基本格式如下。

```
int load(Context context, int resId, int priority)
int load(String path, int priority)
int load(AssetFileDescriptor afd, int priority)
int load(FileDescriptor fd, long offset, long length, int priority)
```

其中，context 为当前应用上下文。resId 是存放到应用的 res/raw 文件中的音频文件的资源 ID。priority 为优先级，目前无效，1 用于与未来版本兼容。path 为存储器中音频文件的路径。AssetFileDescriptor 为音频 asset 文件的描述符。在将多个音频存放在一个二进制文件中时，FileDescriptor 为该音频文件的描述符，offset 指定加载的音频在文件中的开始位置，length 指定音频长度。load()方法返回值为加载的音频的 ID，在调用其他方法进行播放、暂停等操作处理音频时，用音频 ID 作为参数。

如下语句加载了两个 res/raw 文件中的音频文件。

```
int soundId1=sp.load(this,R.raw.winlog,1);
int soundId2=sp.load(this,R.raw.lesson1,1);
```

调用 load()方法，准备好音频资源后，可调用 play()方法来播放音频。play()方法的基本格式如下。

```
play(int soundID, float leftVolume, float rightVolume, int priority, int loop, float rate)
```

soundID 为 load()方法加载音频资源时返回的 ID。leftVolume 和 rightVolume 分别为左声道音量和右声道音量，取值范围为 0.0～1.0。priority 为优先级，0 为最低级。loop 为重复次数，0 表示不重复。rate 为播放速率，取值范围为 0.5～2.0，1.0 为正常播放速度。

如下语句实现了播放 soundID1 对应的音频。

```
sp.play(soundID1,1,1,1,0,1);
```

其他 SoundPool 的常用方法如下。

- pause(int streamID)：暂停播放。
- release()：释放 SoundPool 中加载的音频资源。
- resume(int streamID)：继续播放暂停的音频。
- setLoop(int streamID, int loop)：设置重复播放次数。
- setVolume(int streamID, float leftVolume, float rightVolume)：设置音量。
- stop(int streamID)：停止播放。
- unload(int soundID)：卸载 SoundPool 中的音频资源。

Android 多媒体之 MediaPlayer 播放声音

6.1.2 使用 MediaPlay 播放音频

MediaPlay 类提供了音频和视频播放功能，本小节先介绍如何用其播放音频。MediaPlay 类比 SoundPool 类提供了更多的音频控制功能，支持更多的音频格式。

在使用 MediaPlay 对象处理音频时，音频可处于多种状态，如图 6-1 所示（该图引用自 http://developer.android.com/reference/android/media/MediaPlayer.html）。在调用 MediaPlay 对象方法时，音频可在相应的不同状态之间进行切换。

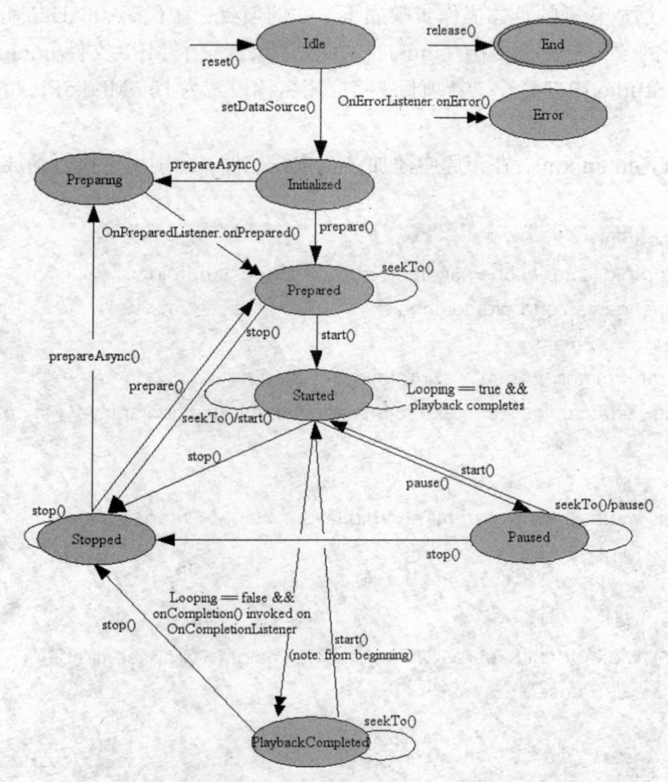

图 6-1 MediaPlay 音频状态

MediaPlay 音频控制的常用方法如下。

- getCurrentPosition：获得当前播放位置。
- getDuration：获得音频时长。
- isPlaying：判断是否处于播放状态。
- pause：暂停播放。
- prepare：准备音频。本地音频通常不需要准备，远程音频通过准备完成下载和本地缓冲。
- release：释放 MediaPlayer 对象资源。
- reset：恢复 MediaPlayer 对象到刚创建时的状态。
- seekTo：设置播放位置。
- setDataSource：设置音频文件位置。
- setVolume：设置音量。
- start：开始播放。
- stop：停止播放。

使用 MediaPlay 播放音频的基本步骤如下。

（1）创建 MediaPlay 对象。

（2）调用 setDataSource 方法设置音频文件路径。

（3）调用 prepare 方法加载音频。

（4）调用 start 方法播放音频。

（5）调用 pause 方法暂停正在播放的音频。

（6）调用 stop 方法停止播放。

（7）调用 reset 方法重置 MediaPlay 到刚创建时的状态。

使用 MediaPlay 播放音频的具体操作步骤如下。实例项目：源代码\06\UseMediaPlay。

（1）在调试设备的 SD 卡根目录下的 music 文件夹中放置一个 MP3 文件 honor.mp3。

（2）在 AndroidStudio 中创建一个新项目，将应用名称设置为 UseMediaPlay，并为项目添加一个空活动。

（3）修改 activity_main.xml，在布局中添加 3 个按钮控件，分别用于执行播放、暂停和停止操作，代码如下。

```
<?xml version="1.0" encoding="utf-8"?>
<LinearLayoutxmlns:android="http://schemas.android.com/apk/res/android"
    xmlns:tools="http://schemas.android.com/tools"
    android:id="@+id/activity_main"
    android:layout_width="match_parent"    android:layout_height="match_parent"
    android:orientation="vertical" tools:context="com.example.xbg.usemediaplayer.MainActivity">
    <Button
        android:text="播放"
        android:layout_width="match_parent" android:layout_height="wrap_content"
        android:id="@+id/btPlayMp3" />
    <Button
        android:text="暂停"
        android:layout_width="match_parent"android:layout_height="wrap_content"
        android:id="@+id/btPauseMp3" />
    <Button
        android:text="停止"
        android:layout_width="match_parent"android:layout_height="wrap_content"
```

```
            android:id="@+id/btStopMp3" />
</LinearLayout>
```

（4）修改 AndroidManifest.xml，添加 SD 卡读取权限，代码如下。

```
<?xml version="1.0" encoding="utf-8"?>
<manifest xmlns:android="http://schemas.android.com/apk/res/android"
    package="com.example.xbg.usemediaplayer">
    <uses-permission android:name="android.permission.READ_EXTERNAL_STORAGE"/>
...
</manifest>
```

（5）修改 MainActivity.java，完成 MediaPlayer 对象的创建，为各个按钮控件添加单击事件监听器，代码如下。

```
package com.example.xbg.usemediaplayer;
import android.Manifest;
...
public class MainActivity extends AppCompatActivity {
    private   MediaPlayer mediaPlayer=null;
    @Override
    protected void onCreate(Bundle savedInstanceState) {
        super.onCreate(savedInstanceState);
        setContentView(R.layout.activity_main);
        mediaPlayer=new MediaPlayer();//创建MediaPlayer对象
        //检查应用是否已经获得授权
        if(ContextCompat.checkSelfPermission(this,
                    Manifest.permission.READ_EXTERNAL_STORAGE)
                    != PackageManager.PERMISSION_GRANTED){
            //如果没有权限，动态申请授权
            ActivityCompat.requestPermissions(this,
                        new String[]{Manifest.permission.READ_EXTERNAL_STORAGE},1);
        }else{
            initMediaPlayer();//初始化MediaPlayer
        }
        Button btPlayMp3= (Button) findViewById(R.id.btPlayMp3);
        btPlayMp3.setOnClickListener(new View.OnClickListener() {
            @Override
            public void onClick(View v) {
                if(!mediaPlayer.isPlaying()){
                    mediaPlayer.start();//开始播放
                }
            }
        });
        Button btPauseMp3= (Button) findViewById(R.id.btPauseMp3);
        btPauseMp3.setOnClickListener(new View.OnClickListener() {
            @Override
            public void onClick(View v) {
                if(mediaPlayer.isPlaying()){
```

```
                mediaPlayer.pause();//暂停播放
            }
        }
    });
    Button btStopMp3= (Button) findViewById(R.id.btStopMp3);
    btStopMp3.setOnClickListener(new View.OnClickListener() {
        @Override
        public void onClick(View v) {
            if(mediaPlayer.isPlaying()){
                mediaPlayer.stop();
                try {
                    mediaPlayer.prepare();
                } catch (IOException e) {
                    e.printStackTrace();
                }
            }
        }
    });
}
private void initMediaPlayer(){
    try {//初始化MediaPlayer
        Filefile=new File(Environment.getExternalStorageDirectory()+"/music","honor.mp3");
        mediaPlayer.setDataSource(file.getPath());//设置音频路径
        mediaPlayer.prepare();//加载音频，完成准备
    } catch (IOException e) {
        e.printStackTrace();
    }
}
@Override
public void onRequestPermissionsResult(int requestCode,
                            @NonNull String[] permissions, @NonNull int[] grantResults) {
    if(requestCode==1){
        if(grantResults.length>0 &&grantResults[0]==
                                    PackageManager.PERMISSION_GRANTED){
            initMediaPlayer();//初始化MediaPlayer
        }else{
            Toast.makeText(this,"未获得SD卡访问权限",Toast.LENGTH_LONG).show();
            finish();
        }
    }
}
@Override
protected void onDestroy() {
    //关闭应用时释放MediaPlayer对象占用的资源
    if(mediaPlayer!=null){
        mediaPlayer.stop();
```

```
                mediaPlayer.release();
                mediaPlayer=null;
            }
            super.onDestroy();
        }
}
```

（6）运行项目，测试运行效果。

应用运行时，如果发现用户还没有授予访问 SD 的权限，则会提示用户，如图 6-2 所示。单击 始终允许 按钮即可授予应用权限。本例中，如果没有为应用授权，则应用会结束运行，因为在没有获得 SD 卡访问权限时访问 SD 卡中的文件会出错。

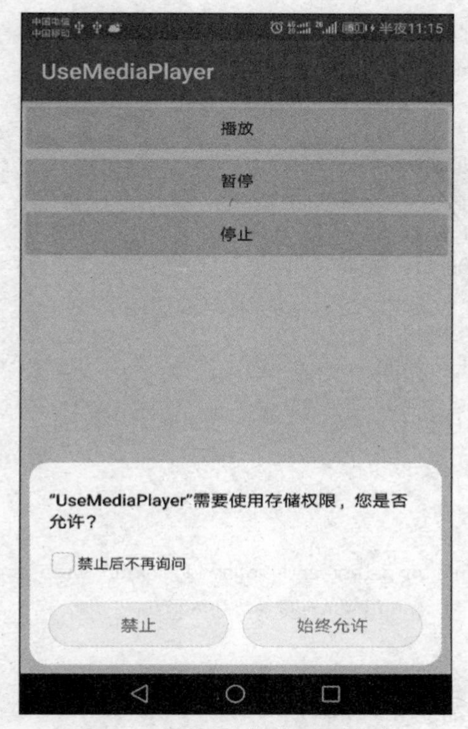

图 6-2　提示用户为应用程序授权

6.1.3　使用 MediaPlayer 播放视频

Android 多媒体之 MediaPlayer 播放视频

MediaPlayer 既可用于播放音频，也可用于播放视频，在用法上没有多大区别。只是在播放视频时应使用 SurfaceView 控件作为视频的显示容器。

使用 MediaPlayer 播放视频的具体操作步骤如下。这里对 6.1.2 小节中的 UseMediaPlay 实例略加修改即可用于播放视频。实例项目：源代码 \06\UseMediaPlay2。

（1）在调试设备的 SD 卡根目录下的 movies 文件夹中添加一个视频文件“广播体操.mp4”。

（2）在 AndroidStudio 中创建一个新项目，将应用名称设置为 UseMediaPlay2，并为项目添加一个空活动。

（3）修改 activity_main.xml，在布局中添加 3 个按钮控件，分别用于执行播放、暂停和停止操作，添加一个 SurfaceView 控件作为视频的显示容器，代码如下。

```
<?xml version="1.0" encoding="utf-8"?>
<LinearLayoutxmlns:android="http://schemas.android.com/apk/res/android"
    xmlns:tools="http://schemas.android.com/tools"
    android:id="@+id/activity_main"
    android:layout_width="match_parent"    android:layout_height="match_parent"
    android:orientation="vertical"    tools:context="com.example.pad.usemediaplayer2.MainActivity">
    <LinearLayout   android:layout_width="match_parent" android:layout_height="wrap_content">
        <Button
            android:text="播放视频"
            android:layout_width="wrap_content"
            android:layout_height="wrap_content"
            android:layout_weight="1"
            android:id="@+id/btPlayVideo" />
        <Button
            android:text="暂停视频"
            android:layout_width="wrap_content"
            android:layout_height="wrap_content"
            android:layout_weight="1"
            android:id="@+id/btPauseVideo" />
        <Button
            android:text="停止视频"
            android:layout_width="wrap_content"
            android:layout_height="wrap_content"
            android:layout_weight="1"
            android:id="@+id/btStopVideo" />
    </LinearLayout>
    <SurfaceView
        android:layout_width="match_parent" android:layout_height="wrap_content"
        android:layout_weight="1"   android:id="@+id/surfaceView" />
</LinearLayout>
```

（4）修改 AndroidManifest.xml，添加 SD 卡读取权限，代码如下。

```
<?xml version="1.0" encoding="utf-8"?>
<manifest xmlns:android="http://schemas.android.com/apk/res/android"
    package="com.example.xbg.usemediaplayer">
    <uses-permission android:name="android.permission.READ_EXTERNAL_STORAGE"/>
…
</manifest>
```

（5）修改 MainActivity.java，代码如下。

```
package com.example.pad.usemediaplayer2;
import android.Manifest;
…
public class MainActivity extends AppCompatActivity {
    private MediaPlayer mediaPlayer=null;
    privateSurfaceView sv;
    privateSurfaceHolder holder;
    privateSurfaceHolder.Callback surfaceHolderCallback=new SurfaceHolder.Callback() {
```

```
        @Override
        public void surfaceCreated(SurfaceHolder holder) {
            mediaPlayer.setDisplay(holder);//将MediaPlayer关联到SurfaceView
        }
        @Override
        public void surfaceChanged(SurfaceHolder holder, int format, int width, int height) {
        }
        @Override
        public void surfaceDestroyed(SurfaceHolder holder) {
        }
    };
    @Override
    protected void onCreate(Bundle savedInstanceState) {
        super.onCreate(savedInstanceState);
        setContentView(R.layout.activity_main);
        mediaPlayer=new MediaPlayer();//创建MediaPlayer对象
        //检查应用是否已经获得授权
        if(ContextCompat.checkSelfPermission(this,
                Manifest.permission.READ_EXTERNAL_STORAGE)
                != PackageManager.PERMISSION_GRANTED){
            //如果没有权限，动态申请授权
            ActivityCompat.requestPermissions(this,
                    new String[]{Manifest.permission.READ_EXTERNAL_STORAGE},1);
        }else{
            initMediaPlayer();//初始化MediaPlayer
        }
        sv= (SurfaceView) findViewById(R.id.surfaceView);
        holder=sv.getHolder();
        holder.addCallback(surfaceHolderCallback);
        Button btPlayVideo= (Button) findViewById(R.id.btPlayVideo);
        btPlayVideo.setOnClickListener(new View.OnClickListener() {
            @Override
            public void onClick(View v) {//开始播放
                mediaPlayer.start();
            }
        });
        Button btPauseVideo= (Button) findViewById(R.id.btPauseVideo);
        btPauseVideo.setOnClickListener(new View.OnClickListener() {
            @Override
            public void onClick(View v) {//暂停播放
                if(mediaPlayer.isPlaying()){
                    mediaPlayer.pause();
                }
            }
        });
        Button btStopVideo= (Button) findViewById(R.id.btStopVideo);
        btStopVideo.setOnClickListener(new View.OnClickListener() {
```

```
                @Override
                public void onClick(View v) {//停止播放
                    mediaPlayer.stop();
                    try {
                        mediaPlayer.prepare();
                    } catch (IOException e) {
                        e.printStackTrace();
                    }
                }
            });
        }
        private void initMediaPlayer(){
            try {//初始化MediaPlayer
                File path = Environment.getExternalStoragePublicDirectory(
                        Environment.DIRECTORY_MOVIES);
                Filefile = new File(path, "广播体操.mp4");
                mediaPlayer.setDataSource(file.getPath());//设置视频路径
                mediaPlayer.prepare();//加载视频，完成准备
            } catch (IOException e) {
                e.printStackTrace();
            }
        }
        @Override
        public void onRequestPermissionsResult(int requestCode,
                                    @NonNull String[] permissions, @NonNull int[] grantResults) {
            if(requestCode==1){
                if(grantResults.length>0 &&grantResults[0]==
                                    PackageManager.PERMISSION_GRANTED){
                    initMediaPlayer();//初始化MediaPlayer
                }else{
                    Toast.makeText(this,"未获得SD卡访问权限",Toast.LENGTH_LONG).show();
                    finish();
                }
            }
        }
        @Override
        protected void onDestroy() {
            //关闭应用时释放MediaPlayer对象占用的资源
            if(mediaPlayer!=null){
                mediaPlayer.stop();
                mediaPlayer.release();
                mediaPlayer=null;
            }
            super.onDestroy();
        }
}
```

（6）运行项目，测试运行效果，如图 6-3 所示。

图 6-3　使用 MediaPlayer 播放视频的效果

6.2　记录声音

MediaRecorder 类提供了音频采集功能，开发者可使用设备的麦克风记录声音。要在应用中实现音频采集功能，首先需要在应用程序的清单文件 AndroidManifest.xml 中添加 RECORD_AUDIO 权限申请使用麦克风，例如如下代码。

Android 多媒体之使用 MediaRecord 录音

```
<uses-permission android:name="android.permission.RECORD_AUDIO" />
```

Android 系统认为使用 RECORD_AUDIO 权限记录用户声音隐私是一种“危险”行为，所以从 Android 6.0（API 23）开始，需要在应用程序运行时动态向用户申请 RECORD_AUDIO 权限。用户授权后，应用可记录授权，不再重复询问。通常调用 ActivityCompat.requestPermissions()方法来动态申请权限。

使用 MediaRecorder 采集音频，并使用 MediaPlayer 播放记录的音频的具体操作步骤如下。实例项目：源代码\06\UseMediaRecorder。

（1）在 AndroidStudio 中创建一个新项目，将应用名称设置为 UseMediaRecorder，并为项目添加一个空活动。

（2）修改 activity_main.xml，为主活动布局按钮控件，代码如下。

```
<?xml version="1.0" encoding="utf-8"?>
<LinearLayoutxmlns:android="http://schemas.android.com/apk/res/android"
    xmlns:tools="http://schemas.android.com/tools"    android:id="@+id/activity_main"
    android:layout_width="match_parent"    android:layout_height="match_parent"
```

```
    android:orientation="vertical" tools:context="com.example.xbg.usemediarecorder.MainActivity">
    <Button   android:text="开始录音" android:layout_width="match_parent"
        android:layout_height="wrap_content"   android:id="@+id/btStartRecord" />
    <Button
        android:text="停止录音"   android:layout_width="match_parent"
        android:layout_height="wrap_content"   android:id="@+id/btStopRecord" />
    <Button
        android:text="播放录音"   android:layout_width="match_parent"
        android:layout_height="wrap_content"   android:id="@+id/btPlay" />
    <Button
        android:text="停止播放"   android:layout_width="match_parent"
        android:layout_height="wrap_content"    android:onClick="stopPlay"
        android:id="@+id/btStop" />
</LinearLayout>
```

（3）修改 MainActivity.java，为各个按钮添加单击事件监听器，实现音频的采集和播放控制，代码如下。

```
package com.example.xbg.usemediarecorder;
import android.Manifest;
...
public class MainActivity extends AppCompatActivity {
    private MediaPlayer mediaPlayer=null;
    privateMediaRecordermediaRecorder=null;
    privateString mFileName = null;
    private static final String LOG_TAG = "UseMediaRecorder";
    @Override
    protected void onCreate(Bundle savedInstanceState) {
        super.onCreate(savedInstanceState);
        setContentView(R.layout.activity_main);
        mFileName = getExternalCacheDir().getAbsolutePath();
        mFileName += "/audiorecord.3gp";
        //检查应用是否已经获得授权
        if(ContextCompat.checkSelfPermission(this,
                Manifest.permission.RECORD_AUDIO)
                != PackageManager.PERMISSION_GRANTED){
            //如果没有权限，动态申请授权
            ActivityCompat.requestPermissions(this,
                    new String[]{Manifest.permission.RECORD_AUDIO},1);
        }else {
            initMediaRecorder();
        }
        Button btStartRecord= (Button) findViewById(R.id.btStartRecord);
        btStartRecord.setOnClickListener(new View.OnClickListener() {
            @Override
            public void onClick(View v) {//开始录音
                try {
                    mediaRecorder.prepare();//准备MediaRecorder
```

```
                } catch (IOException e) {
                    Log.e(LOG_TAG, "准备MediaRecorder出错啦！");
                }
                mediaRecorder.start();//开始采集音频
            }
        });
        Button btStopRecord= (Button) findViewById(R.id.btStopRecord);
        btStopRecord.setOnClickListener(new View.OnClickListener() {
            @Override
            public void onClick(View v) {//停止录音
                mediaRecorder.stop();//停止MediaRecorder
                mediaRecorder.release();//释放MediaRecorder所占资源
                mediaRecorder = null;
            }
        });
        Button btStartPlay= (Button) findViewById(R.id.btPlay);
        btStartPlay.setOnClickListener(new View.OnClickListener() {
            @Override
            public void onClick(View v) {//开始播放
                mediaPlayer = new MediaPlayer();
                try {
                    mediaPlayer.setDataSource(mFileName);//设置要播放的音频文件
                    mediaPlayer.prepare();
                    mediaPlayer.start();
                } catch (IOException e) {
                    Log.e(LOG_TAG, "MediaPlayer方法prepare()执行失败！");
                }
            }
        });
        Button btStop= (Button) findViewById(R.id.btStop);
        btStop.setOnClickListener(new View.OnClickListener() {
            @Override
            public void onClick(View v) {//停止播放
                mediaPlayer.release();
                mediaPlayer = null;
            }
        });
    }
    @Override
    public void onRequestPermissionsResult(int requestCode,
                                  @NonNull String[] permissions, @NonNull int[] grantResults) {
        if(requestCode==1){
            if(grantResults.length>0 &&grantResults[0]==
                    PackageManager.PERMISSION_GRANTED){
                initMediaRecorder();//初始化
            }else{
                Toast.makeText(this,"未获得麦克风访问权限", Toast.LENGTH_LONG).show();
```

```
                finish();
            }
        }
    }
    private void initMediaRecorder() {//初始化MediaRecorder
        mediaRecorder = new MediaRecorder();
        mediaRecorder.setAudioSource(MediaRecorder.AudioSource.MIC);//设置音频来源，使用麦克风
        mediaRecorder.setOutputFormat(MediaRecorder.OutputFormat.THREE_GPP);//设置输出格式
        mediaRecorder.setOutputFile(mFileName);//设置音频输出文件
        mediaRecorder.setAudioEncoder(MediaRecorder.AudioEncoder.AMR_NB);//设置音频编码方式
    }
    @Override
    protected void onDestroy() {//应用停止时，释放资源
        if (mediaRecorder != null) {
            mediaRecorder.release();
            mediaRecorder = null;
        }
        if (mediaPlayer != null) {
            mediaPlayer.release();
            mediaPlayer = null;
        }
        super.onDestroy();
    }
}
```

（4）运行项目，测试运行效果。程序运行时，首先会提示应用要使用麦克风，需要用户授权，如图 6-4 所示。只有在用户允许后，应用程序才能继续运行。

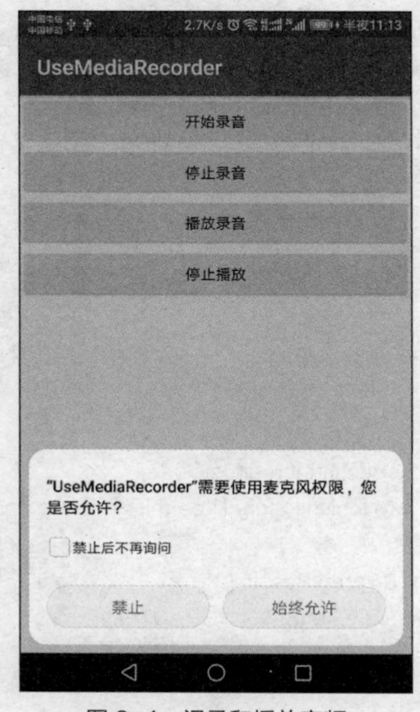

图 6-4 记录和播放音频

单击 开始录音 按钮，即可通过麦克风记录声音。单击 停止录音 按钮，即停止记录。单击 停止录音 按钮，即可播放记录下来的声音。

6.3 使用摄像头和相册

在常用通信软件（如QQ、微信等）中，经常需要分享图片，这些图片可来自于相册或者摄像头拍摄。

Android多媒体之使用系统照相机拍照

6.3.1 使用摄像头拍摄照片

要使用摄像头拍摄照片，最简单的方式就是调用系统拍照程序。创建一个动作为MediaStore.ACTION_IMAGE_CAPTURE的Intent对象，执行startActivityForResult()方法启动系统拍照程序，即可使用摄像头进行拍照。

调用系统拍照程序完成拍照，并显示拍摄的照片的具体操作步骤如下。实例项目：源代码\06\UseSystemCameraApp。

（1）在AndroidStudio中创建一个新项目，将应用名称设置为UseSystemCameraApp，并为项目添加一个空活动。

（2）修改activity_main.xml，为主活动布局一个Button控件和一个ImageView控件，Button控件用于启动系统拍照程序，ImageView控件用于显示所拍照片，代码如下。

```
<?xml version="1.0" encoding="utf-8"?>
<LinearLayoutxmlns:android="http://schemas.android.com/apk/res/android"
    xmlns:tools="http://schemas.android.com/tools"
    android:id="@+id/activity_main"
    android:layout_width="match_parent"
    android:layout_height="match_parent"
    android:orientation="vertical"
    tools:context="com.example.xbg.usesystemcameraapp.MainActivity">
    <Button
        android:text="使用系统照相应用"
        android:layout_width="match_parent"
        android:layout_height="wrap_content"
        android:id="@+id/btTakePhoto" />
    <ImageView
        android:id="@+id/ivShow"
        android:layout_weight="1"
        android:layout_width="match_parent"
        android:layout_height="wrap_content"/>
</LinearLayout>
```

（3）由于需要将所拍照片保存到设备公共的 PICTURES 目录中，所以需要在应用程序清单文件AndroidManifest.xml中申请权限，代码如下。

```
<?xml version="1.0" encoding="utf-8"?>
<manifest xmlns:android="http://schemas.android.com/apk/res/android"
    package="com.example.xbg.usesystemcameraapp">
    <uses-permission android:name="android.permission.WRITE_EXTERNAL_STORAGE"/>
    ...
</manifest>
```

（4）修改 MainActivity.java，实现拍照和照片显示，代码如下。

```
package com.example.xbg.usesystemcameraapp;
import android.content.Intent;
...
public class MainActivity extends AppCompatActivity {
    privateFilepicFile=null;
    @Override
    protected void onCreate(Bundle savedInstanceState) {
        super.onCreate(savedInstanceState);
        setContentView(R.layout.activity_main);
        //使用StrictMode.VmPolicy.Builder监测应用中的FileUriExposure事件
        StrictMode.VmPolicy.Builder builder = new StrictMode.VmPolicy.Builder();
        StrictMode.setVmPolicy(builder.build());
        builder.detectFileUriExposure();
        Button btTakePhoto= (Button) findViewById(R.id.btTakePhoto);
        btTakePhoto.setOnClickListener(new View.OnClickListener() {
            @Override
            public void onClick(View v) {
                try {
                    //创建用于保存所拍照片的文件
                    Filesdcard = Environment.getExternalStoragePublicDirectory(
                                                Environment.DIRECTORY_PICTURES);
                    picFile = new File(sdcard, System.currentTimeMillis() + ".jpg");
                    picFile.createNewFile();
                    Log.e("UseSystemCameraApp",picFile.getName()+"创建成功！ ");
                }catch(IOException e){
                    e.printStackTrace();
                }
                //使用Intent调用系统摄像头拍照程序
                Intent intent=new Intent(MediaStore.ACTION_IMAGE_CAPTURE);
                intent.putExtra(MediaStore.EXTRA_OUTPUT, Uri.fromFile(picFile));
                startActivityForResult(intent,1);
            }
        });
    }

    @Override
    protected void onActivityResult(int requestCode, int resultCode, Intent data) {
        if(requestCode==1){
            //处理ActivityResult调用返回，将所拍照片显示在ImageView中
            ImageView iv=(ImageView)findViewById(R.id.ivShow);
            iv.setImageURI(Uri.fromFile(picFile));
        }
        super.onActivityResult(requestCode, resultCode, data);
    }
}
```

（5）运行项目，测试运行效果。程序运行效果如图 6-5 所示，ImageView 控件中显示了所拍照片。

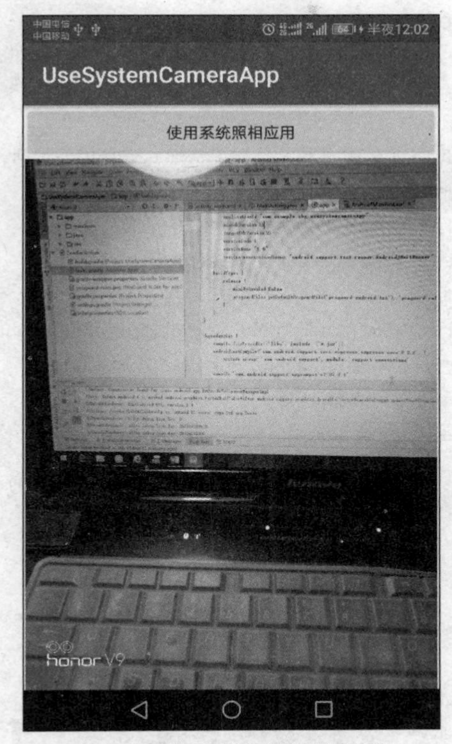

图 6-5　使用系统拍照程序拍照

本例使用了如下语句将准备好用于保存照片的文件 URI 包含到 Intent 中，并传递给系统拍照程序。

```
intent.putExtra(MediaStore.EXTRA_OUTPUT, Uri.fromFile(picFile));
```

这意味着将当前 Activity 中的文件 URI 暴露给了另一个 Activity，如果不进行处理，程序运行时会报错。所以在 onCreate()方法中，须调用 StrictMode.VmPolicy.Builder 的 detectFileUriExposure()方法监测文件 URI 暴露信息，避免程序出错。

6.3.2　选取相册图片

选取相册图片与使用系统拍照程序拍照类似，使用 Android 内置的 Activity 即可完成。将 6.3.1 小节中的实例项目 UseSystemCameraApp 略加修改，即可实现相册图片选择，步骤如下。实例项目：源代码\06\SelectPhoto。

（1）创建 Intent 对象，指定 Intent.ACTION_PICK 操作用于启动相册，代码如下。

```
Button btSelectPhoto= (Button) findViewById(R.id.btSelectPhoto);
btSelectPhoto.setOnClickListener(new View.OnClickListener() {
    @Override
    public void onClick(View v) {
        //使用Intent对象来打开相册
        Intent intent = new Intent(Intent.ACTION_PICK,
                android.provider.MediaStore.Images.Media.EXTERNAL_CONTENT_URI);
        startActivityForResult(intent, 1);
    }
});
```

（2）在 onActivityResult()方法中处理返回的图片 URI，将图片显示到 ImageView 中，代码如下。

```
@Override
    protected void onActivityResult(int requestCode, int resultCode, Intent data) {
        super.onActivityResult(requestCode, resultCode, data);
        if(requestCode==1){//处理从相册选取照片的返回结果
            if(resultCode==RESULT_OK){//若用户正确完成照片选择操作返回，则进一步处理选择的照片
                Uri uri = data.getData();
                Log.e("图片URI：", uri.toString());
                ContentResolver cr = this.getContentResolver();
                try {
                    Bitmap bitmap = BitmapFactory.decodeStream(cr.openInputStream(uri));
                    ImageView iv=(ImageView)findViewById(R.id.ivShow);
                    /* 将Bitmap设定到ImageView */
                    iv.setImageBitmap(bitmap);
                } catch (FileNotFoundException e) {
                    Log.e("出错了：", e.getMessage(),e);
                }
            }
        }
    }
```

程序运行效果如图 6-6 所示。

图 6-6　选取相册图片

6.4 编程实践：自定义音乐播放器

本节将综合应用本章所学知识，实现一个音乐播放器，程序运行效果如图 6-7 所示。屏幕下方的列表中显示了当前设备中的所有 MP3 乐曲文件信息，单击某一首乐曲，可开始播放。屏幕上方显示了当前播放音乐的名称和播放进度，并可通过拖动播放进度条改变播放位置，或者通过按钮操作播放、暂停或停止当前音乐。

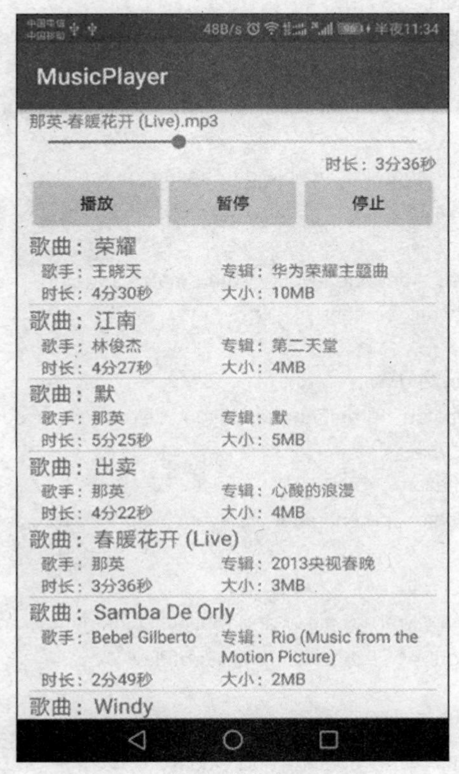

图 6-7 自定义的音乐播放器

本节实例将使用 MediaPlayer 来完成乐曲的播放控制，其使用方法在本章前面的内容中已经讲解。本节需要用到的新知识是 Android 的媒体库 MediaStore。媒体库是 Android 自动为设备中的音频、视频、图片以及其他非多媒体文件等资源的元信息创建的 SQLite 数据库。

Android 主要通过下述 4 个静态类来访问各种文件的元信息。

- MediaStore.Audio：存放所有音频文件的信息。
- MediaStore.Files：存放所有多媒体文件和非多媒体文件（如.pdf、.html、.txt 等）的信息。
- MediaStore.Images：存放所有图片文件的信息。
- MediaStore.Video：存放所有视频文件的信息。

媒体库中的所有信息通过 ContentProvider 的数据共享机制进行共享，系统中的各种应用均可访问。

本节实例将访问 MediaStore.Audio 获取设备中的 MP3 文件信息，然后进行播放。MediaStore.Audio 中存放的音频元信息包括歌曲名称（Title）、歌手（Artist）、专辑（Album）、时长（Duration）、文件大小（size）、文件名（DisplayName）及文件路径（Data）等信息。在歌曲列表中单击歌曲名称，即可将歌曲的文件路径传递给 MediaPlayer 并进行播放。本节实例的具体操作步骤如下。

（1）在 AndroidStudio 中创建一个新项目，将应用名称设置为 MusicPlayer，并为项目添加一个空活动。

（2）修改 activity_main.xml，在主活动布局中添加各种控件，代码如下。

```
<?xml version="1.0" encoding="utf-8"?>
<LinearLayoutxmlns:android="http://schemas.android.com/apk/res/android"
    xmlns:tools="http://schemas.android.com/tools"
    android:id="@+id/activity_main"
    android:layout_width="match_parent"   android:layout_height="match_parent"
    android:layout_marginLeft="10dp"   android:layout_marginRight="10dp"
    android:orientation="vertical" tools:context="com.example.xbg.musicplayer.MainActivity">
    <TextView
        android:text="当前播放：" android:layout_width="match_parent"
        android:layout_height="wrap_content" android:id="@+id/tvName" />
    <SeekBar
        android:id="@+id/sbSeek"   android:layout_width="match_parent"
        android:layout_height="wrap_content" />
    <TextView
        android:text="时长" android:gravity="right"
        android:layout_width="match_parent" android:layout_height="wrap_content"
        android:id="@+id/tvLen" />
    <LinearLayout android:orientation="horizontal"
        android:layout_width="match_parent" android:layout_height="wrap_content">
        <Button
            android:text="播放"
            android:layout_width="wrap_content" android:layout_height="wrap_content"
            android:layout_weight="1"   android:id="@+id/btPlay" />
        <Button
            android:text="暂停"
            android:layout_width="wrap_content" android:layout_height="wrap_content"
            android:layout_weight="1" android:id="@+id/btPause" />
        <Button
            android:text="停止"
            android:layout_width="wrap_content" android:layout_weight="1"
            android:layout_height="wrap_content" android:id="@+id/btStop" />
        <Button
            android:text="刷新"
            android:layout_width="wrap_content" android:layout_weight="1"
            android:layout_height="wrap_content" android:id="@+id/btRefresh" />
    </LinearLayout>
    <ListView
        android:id="@+id/lsAudios"
        android:layout_weight="1" android:layout_width="match_parent"
        android:layout_height="wrap_content" />
</LinearLayout>
```

（3）添加一个布局文件 songitem.xml，定义歌曲列表中每首歌曲的信息显示布局，代码如下。

```
<?xml version="1.0" encoding="utf-8"?>
<LinearLayoutxmlns:android="http://schemas.android.com/apk/res/android"
    xmlns:tools="http://schemas.android.com/tools"
    android:orientation="vertical" android:layout_width="match_parent"
    android:layout_height="match_parent" android:layout_marginLeft="10dp"
    android:layout_marginRight="10dp">
    <TextView
        android:text="TextView"
        android:layout_width="match_parent"   android:layout_height="wrap_content"
        android:id="@+id/tvTitle" android:textSize="18sp" />
    <TableLayout
        android:layout_width="match_parent"   android:layout_height="match_parent"
        android:layout_marginLeft="10dp"   android:orientation="vertical">
        <TableRowandroid:layout_width="match_parent" android:layout_height="match_parent" >
            <TextView
                android:text="Singer" android:layout_width="150dp"
                android:layout_height="wrap_content"   android:id="@+id/tvSinger" />
            <TextView
                android:text="Album"
                android:layout_width="wrap_content"   android:layout_height="wrap_content"
                android:id="@+id/tvAlbum"   android:layout_weight="1" />
        </TableRow>
        <TableRow android:layout_width="match_parent"   android:layout_height="match_parent">
            <TextView
                android:text="Duration"   android:layout_width="150dp"
                android:layout_height="wrap_content"   android:id="@+id/tvDuration"/>
            <TextView
                android:text="Size"   android:layout_width="wrap_content"
                android:layout_height="wrap_content" android:id="@+id/tvSize"
                android:layout_weight="1" />
        </TableRow>
    </TableLayout>
</LinearLayout>
```

（4）添加一个 Java 类，封装歌曲信息，代码如下。

```
package com.example.xbg.musicplayer;
public class Song {
    private String fileName;
    private String title;
    private int duration;
    private String singer;
    private String album;
    private String size;
    private String filePath;

    public Song(String fileName, String title, int duration, String singer,
```

```
                    String album, String size, String filePath) {
        this.fileName = fileName;
        this.title = title;
        this.duration = duration;
        this.singer = singer;
        this.album = album;
        this.size = size;
        this.filePath = filePath;
    }
    public void setFileName(String fileName) {this.fileName = fileName; }
    public void setTitle(String title) { this.title = title; }
    public void setDuration(int duration) { this.duration = duration; }
    public void setSinger(String singer) { this.singer = singer;      }

    public void setAlbum(String album) { this.album = album;      }
    public void setSize(String size) { this.size = size;     }
    public void setFilePath(String filePath) {this.filePath = filePath;      }
    public String getFileName() { return fileName;      }
    public String getTitle() { return title;      }
    public int getDuration() { return duration;      }
    public String getSinger() {return singer;      }
    public String getAlbum() { return album;      }
    public String getSize() {return size;      }
    public String getFilePath() {return filePath;      }
}
```

（5）添加一个 Java 类，实现自定义的适配器填充 ListView，代码如下。

```
package com.example.xbg.musicplayer;
import android.content.Context;
...
public class SongAdapter extends ArrayAdapter<Song> {
    public int resId;
    publicSongAdapter(Context context, int resource, List<Song> objects) {
        super(context, resource, objects);
        resId=resource;
    }
    @NonNull
    @Override
    public View getView(int position, View convertView, ViewGroup parent) {
        Song song=getItem(position);
        View view= LayoutInflater.from(getContext()).inflate(resId,parent,false);
        TextViewtvTile= (TextView) view.findViewById(R.id.tvTitle);
        TextViewtvSinger= (TextView) view.findViewById(R.id.tvSinger);
        TextViewtvAlbum= (TextView) view.findViewById(R.id.tvAlbum);
        TextViewtvDuration= (TextView) view.findViewById(R.id.tvDuration);
        TextViewtvSize= (TextView) view.findViewById(R.id.tvSize);
```

```
            tvTile.setText("歌曲："+song.getTitle());
            tvSinger.setText("歌手："+song.getSinger());
            tvAlbum.setText("专辑："+song.getAlbum());
            int m=song.getDuration()/60000;
            int s=(song.getDuration()-m*60000)/1000;
            tvDuration.setText("时长："+ m+"分"+s+"秒");
            tvSize.setText("大小："+song.getSize());
            return view;
        }
}
```

（6）修改 MainActivity.java，代码如下。

```
package com.example.xbg.musicplayer;
import android.Manifest;
...
public class MainActivity extends AppCompatActivity {
    private MediaPlayer mediaPlayer=null;
    private static int SELECT_MUSIC=100;
    privateTextViewtvMusicName;
    privateSeekBarsbSeek;
    privateTextViewtvLen;
    private Uri musicUri;
    privateList<Song>listSong;
    private Timer mTimer;
    privateTimerTaskmTimerTask;
    private boolean seekBarChange=false;
    @Override
    protected void onCreate(Bundle savedInstanceState) {
        super.onCreate(savedInstanceState);
        setContentView(R.layout.activity_main);
        //检查应用是否已经获得授权
        if(ContextCompat.checkSelfPermission(this,
                    Manifest.permission.READ_EXTERNAL_STORAGE)
                    != PackageManager.PERMISSION_GRANTED){
            //如果没有权限，则动态申请授权
            ActivityCompat.requestPermissions(this,
                        new String[]{Manifest.permission.READ_EXTERNAL_STORAGE},1);
        }
        initAudioList();//初始化歌曲列表
        tvMusicName= (TextView) findViewById(R.id.tvName);
        tvLen= (TextView) findViewById(R.id.tvLen);
        sbSeek = (SeekBar) findViewById(R.id.sbSeek);
        sbSeek.setOnSeekBarChangeListener(new SeekBar.OnSeekBarChangeListener() {
            @Override
            public void onProgressChanged(SeekBarseekBar, int progress, boolean fromUser) {}
            @Override
```

```
        public void onStartTrackingTouch(SeekBarseekBar) {
            seekBarChange=true;
        }
        @Override
        public void onStopTrackingTouch(SeekBarseekBar) {
            seekBarChange=false;
            mediaPlayer.seekTo(seekBar.getProgress());//拖动SeekBar进度时，改变歌曲播放进度
        }
    });
    Button btPlay= (Button) findViewById(R.id.btPlay);
    btPlay.setOnClickListener(new View.OnClickListener() {
        @Override
        public void onClick(View v) {
            if(mediaPlayer!=null)   mediaPlayer.start();//开始播放
        }
    });
    Button btPause= (Button) findViewById(R.id.btPause);
    btPause.setOnClickListener(new View.OnClickListener() {
        @Override
        public void onClick(View v) {
            if(mediaPlayer!=null && mediaPlayer.isPlaying()){
                mediaPlayer.pause();//暂停播放
            }
        }
    });
    Button btStop= (Button) findViewById(R.id.btStop);
    btStop.setOnClickListener(new View.OnClickListener() {
        @Override
        public void onClick(View v) {//停止播放
            if(mediaPlayer!=null && mediaPlayer.isPlaying()){
                mediaPlayer.stop();
                try {
                    mediaPlayer.prepare();
                } catch (IOException e) {
                    e.printStackTrace();
                }
            }
        }
    });
    Button btRefresh= (Button) findViewById(R.id.btRefresh);
    btRefresh.setOnClickListener(new View.OnClickListener() {
        @Override
        public void onClick(View v) {//刷新歌曲列表
            initAudioList();
        }
```

```
        });

        ListView lsAudios= (ListView) findViewById(R.id.lsAudios);
        lsAudios.setOnItemClickListener(new AdapterView.OnItemClickListener() {
            @Override
            public void onItemClick(AdapterView<?> parent, View view, int position, long id) {
                Song song=listSong.get(position);
                initMediaPlayer(song);//在歌曲列表中单击歌曲名称时播放该歌曲
            }
        });
    }
    private void initAudioList() {//从媒体库获取MP3信息，填充到歌曲列表中
        //刷新媒体库
        MediaScannerConnection.scanFile(this, new String[] { Environment
                .getExternalStorageDirectory().getAbsolutePath() }, null, null);
        //查询媒体库，用获得的乐曲信息填充ListView
        listSong=new ArrayList<Song>();
        //从媒体库查询MP3类型的乐曲信息
        Cursor musics = getContentResolver().query(
                MediaStore.Audio.Media.EXTERNAL_CONTENT_URI,
                new String[] { MediaStore.Audio.Media._ID,
                        MediaStore.Audio.Media.DISPLAY_NAME,
                        MediaStore.Audio.Media.TITLE,
                        MediaStore.Audio.Media.DURATION,
                        MediaStore.Audio.Media.ARTIST,
                        MediaStore.Audio.Media.ALBUM,
                        MediaStore.Audio.Media.SIZE,
                        MediaStore.Audio.Media.DATA },
                MediaStore.Audio.Media.MIME_TYPE + "=?",
                new String[] { "audio/mpeg"},null);
        String fileName,title,singer,album,year,size,filePath="";
        int duration,m,s;
        Song song;
        if(musics.moveToFirst())
            do{
                fileName=musics.getString(1);
                title=musics.getString(2);
                duration=musics.getInt(3);
                singer=musics.getString(4);
                album=musics.getString(5);
                size=(musics.getString(6)==null)?"未知":musics.getInt(6)/1024/1024+"MB";
                if(musics.getString(7)!=null)    filePath=musics.getString(7);
                song=new Song(fileName,title,duration,singer,album,size,filePath);
                listSong.add(song);
            }while(musics.moveToNext());
        musics.close();
```

```
        SongAdapter adapter=new SongAdapter(MainActivity.this,R.layout.songitem,listSong);
        ListView listView=(ListView)findViewById(R.id.lsAudios);
        listView.setAdapter(adapter);
    }
    private void initMediaPlayer(Song song){
        try {//初始化MediaPlayer
            if(mediaPlayer==null)
                mediaPlayer=new MediaPlayer();//创建MediaPlayer对象
            mediaPlayer.reset();
            mediaPlayer.setDataSource(song.getFilePath());//设置音频路径
            mediaPlayer.prepare();//加载音频，完成准备
            mediaPlayer.start();
            int m=song.getDuration()/60000;
            int s=(song.getDuration()-m*60000)/1000;
            tvLen.setText("时长："+ m+"分"+s+"秒");
            tvMusicName.setText(song.getFileName());
            sbSeek.setMax(song.getDuration());
            //使用定时器记录播放进度，并实时更新SeekBar进度
            mTimer = new Timer();
            mTimerTask = new TimerTask() {
                @Override
                public void run() {
                    if(seekBarChange) return;
                    sbSeek.setProgress(mediaPlayer.getCurrentPosition());

                }
            };
            mTimer.schedule(mTimerTask, 0, 10);
        } catch (Exception e) {
            e.printStackTrace();
        }
    }
        @Override
    public void onRequestPermissionsResult(int requestCode,
                                @NonNull String[] permissions, @NonNull int[] grantResults) {
        if(requestCode==1){
            if(!(grantResults.length>0 &&grantResults[0]==
                    PackageManager.PERMISSION_GRANTED)){
                Toast.makeText(this,"未获得SD卡访问权限",Toast.LENGTH_LONG).show();
                finish();
            }
        }
    }
    @Override
    protected void onDestroy() {
        //关闭应用时释放MediaPlayer对象占用的资源
```

```
        if(mediaPlayer!=null){
            mediaPlayer.stop();
            mediaPlayer.release();
            mediaPlayer=null;
        }
        super.onDestroy();
    }
}
```

（7）运行项目，测试运行效果。

6.5 小结

本章主要介绍了 Android 中基本的多媒体处理功能，包括播放多媒体文件、记录声音、使用摄像头和相册等。SoundPool 一般适用于播放较短的音效。MediaPlayer 既可播放音频，也可播放视频。使用 MediaRecorder 记录声音时，需要访问麦克风，此时需要在应用中动态向用户申请权限。在使用摄像头时，最简单的是直接调用 Android 内置的拍照程序进行拍照，也可用最新的 Camera2 API 来实现自定义的拍照程序。

另外，本章实例中用到了 Android 媒体库的知识，可通过访问媒体库来选取相册图片和获取歌曲信息。媒体库为开发人员提供了获取存储器中各种文件元信息的便捷方式，感兴趣的读者可查看相关资料来了解详细内容。

6.6 习题

1. 简述使用 SoundPool 播放音频的基本步骤。
2. 简述使用 MediaPlayer 播放音频的基本步骤。
3. 简述使用 MediaRecorder 记录声音的基本步骤。
4. 简述使用摄像头拍照的基本步骤。

PART07

第7章 网络和数据解析

- 使用WebView
- 基于HTTP的网络访问方法
- 解析XML格式数据
- 解析JSON数据

■ 随着信息技术的发展，人们生活中的各种设备，无论是 PC、手机、平板电脑，还是电视，甚至冰箱、洗衣机等各种家用电器，都逐渐具备了网络访问功能。Android 设备毫无疑问也是可以上网的，其内置的 QQ、微信、摩拜单车等常见应用都使用了网络技术。本章将介绍如何在 Android 系统中使用 HTTP 与网络服务器进行通信，以获取服务器数据进行解析。

7.1 使用WebView

WebView 控件用于在 Android 应用中代替浏览器来显示网页。使用 WebView 显示网页的具体操作步骤如下。实例项目：源代码\07\UseWebView。

（1）在 AndroidStudio 中创建一个新项目，将应用名称设置为 UseWebView，并为项目添加一个空活动。

（2）修改应用程序清单文件 AndroidManifest.xml，申明网络访问权限，代码如下。

```
<?xml version="1.0" encoding="utf-8"?>
<manifest xmlns:android="http://schemas.android.com/apk/res/android"
    package="com.example.xbg.usewebview">
    <uses-permission android:name="android.permission.INTERNET"/>
    ...
</manifest>
```

（3）修改 activity_main.xml，在主活动布局中添加一个 WebView 控件，代码如下。

```
<?xml version="1.0" encoding="utf-8"?>
<RelativeLayoutxmlns:android="http://schemas,android.com/apk/res/android"
    xmlns:tools="http://schemas.android.com/tools"
    android:id="@+id/activity_main"
    android:layout_width="match_parent"
    android:layout_height="match_parent"
    android:paddingBottom="@dimen/activity_vertical_margin"
    android:paddingLeft="@dimen/activity_horizontal_margin"
    android:paddingRight="@dimen/activity_horizontal_margin"
    android:paddingTop="@dimen/activity_vertical_margin"
    tools:context="com.example.xbg.usewebview.MainActivity">
    <WebView
        android:layout_width="match_parent"
        android:layout_height="match_parent"
        android:id="@+id/webView" />
</RelativeLayout>
```

（4）修改 MainActivity.java，代码如下。

```
package com.example.xbg.usewebview;
import android.support.v7.app.AppCompatActivity;
...
public class MainActivity extends AppCompatActivity {
    @Override
    protected void onCreate(Bundle savedInstanceState) {
        super.onCreate(savedInstanceState);
        setContentView(R.layout.activity_main);
        WebViewwebView= (WebView) findViewById(R.id.webView);
        WebSettings ws=webView.getSettings();
        ws.setJavaScriptEnabled(true);                        //启用JavaScript
        webView.setWebViewClient(new WebViewClient());           //使页面导航保持在WebView中
```

```
        webView.loadUrl("http://developer.android.google.cn");        //载入网页
    }
}
```

（5）运行项目，测试运行效果。运行效果如图 7-1 所示。

图 7-1　使用 WebView 显示网页的效果

WebView 控件的使用很简单，首先通过 WebSettings 对象启用 JavaScript 支持，目前绝大多数网页都使用了 JavaScript 脚本技术。本例中只设置了启用 JavaScript，实际应用中可调用类似方法进行其他设置。

setWebViewClient()方法为 WebView 控件设置了一个 WebViewClient 对象，当用户在页面中单击超链接时，仍然在当前 WebView 中显示链接的页面，否则 Android 会再次打开系统浏览器来显示链接页面。

7.2　基于 HTTP 的网络访问方法

HTTP 是一种超文本传输协议，上网浏览网页时的数据传输必须遵循 HTTP。简单地说，用户在浏览器中打开一个网页地址时，意味着从客户端向服务器发送了一个请求，服务器接收到该请求后进行相应处理，然后将处理结果返回客户端，浏览器解析返回的结果，将最终的解析结果展示给用户。

WebView 控件封装了发起客户端请求、接收服务器响应、解析响应和显示结果等 HTTP 的客户端所有操作。当需要直接获得服务器响应结果时，可使用 HttpURLConnection 等类似方法。

7.2.1 使用 HttpURLConnection

早期的 Android 提供了两种 HTTP 请求的方法，即 HttpClient API 和 HttpURLConnection。HttpClient API 使用起来比较复杂，已经在 Android 6.0 中被移除。HttpURLConnection 比较简单，被推荐继续使用。HttpURLConnection 用于发起 HTTP 请求，并获得返回结果。

使用 HttpURLConnection 的基本步骤如下。

（1）调用 URL 对象的 openConnection()方法获得 HttpURLConnection 实例对象，代码如下。

```
URL url=new URL("https://developer.android.google.cn ");
HttpURLConnection con=(HttpURLConnection)url.openConnection();
```

（2）设置 HTTP 请求方法，代码如下。

```
con.setRequestMethod("GET");
```

常用的 HTTP 请求方法主要有 GET 和 POST 两种（注意大写）。GET 方法一般用于仅希望从服务器返回数据，POST 则可向服务器提交数据。

（3）设置请求相关参数，例如，可设置连接和请求的超时时间（单位为 ms），代码如下。

```
con.setConnectTimeout(6000);
con.setReadTimeout(6000);
```

如果采用 POST 方式，则需要使用 DataOutputStream 来添加需要向服务器提交的数据，代码如下。

```
con.setRequestMethod("POST");
con.setDoOutput(true);
DataOutputStream out=new DataOutputStream(con.getOutputStream());
out.writeBytes("id=admin&pwd=123");
```

向服务器提交的数据使用键值对的方式表示，键值对之间用&符号分隔。

（4）处理返回结果。调用 HttpURLConnection 对象的 getInputStream()方法获得服务器返回结果的 InputStream，从中可获取服务器返回结果，代码如下。

```
InputStream in=con.getInputStream();
reader =new BufferedReader(new InputStreamReader(in));
StringBuilder result=new StringBuilder();
String s;
s=reader.readLine();
while(s!=null){
    result.append(s);
    s=reader.readLine();
}
```

上述代码可将 InputStream 中的服务器返回结果读取到一个 StringBuilder 对象中，再调用其 toString()方法即可获得返回结果的字符串形式。

下面通过一个具体的实例说明如何使用 HttpURLConnection 完成 HTTP 请求，具体操作步骤如下。实例项目：源代码\07\UseHttpURLConnection。

（1）在 AndroidStudio 中创建一个新项目，将应用名称设置为 UseHttpURLConnection，并为项目添加一个空活动。

（2）修改应用程序清单文件 AndroidManifest.xml，申明网络访问权限。

（3）修改 activity_main.xml，在主活动布局中添加一个 TextView 控件，代码如下。

```
<?xml version="1.0" encoding="utf-8"?>
```

```
<RelativeLayoutxmlns:android="http://schemas.android.com/apk/res/android"
    ...
    tools:context="com.example.xbg.usehttpurlconnection.MainActivity">
    <TextView
        android:layout_width="match_parent"
        android:layout_height="match_parent"
        android:id="@+id/textView" />
</RelativeLayout>
```

（4）修改 MainActivity.java，添加按钮的单击事件监听器。在单击按钮时，首先创建 Intent 对象，然后在其中封装数据，最后用其启动活动，代码如下。

```
package com.example.xbg.usehttpurlconnection;
import android.support.v7.app.AppCompatActivity;
...
public class MainActivity extends AppCompatActivity {
    TextViewtextView;
    @Override
    protected void onCreate(Bundle savedInstanceState) {
        super.onCreate(savedInstanceState);
        setContentView(R.layout.activity_main);
        textView= (TextView) findViewById(R.id.textView);
        doUrlGet();//执行HTTP请求
    }
    private void doUrlGet(){//完成HTTP请求
        HttpURLConnection con=null;
        BufferedReader reader=null;
        try {
            URL url=new URL("https://developer.android.google.cn");
            con=(HttpURLConnection)url.openConnection();
            con.setRequestMethod("GET");
            InputStream in=con.getInputStream();
            reader =new BufferedReader(new InputStreamReader(in));
            StringBuilder result=new StringBuilder();
            String s;
            s=reader.readLine();
            while(s!=null){
                result.append(s);
                s=reader.readLine();
            }
            showResult(result.toString());
        } catch (Exception e) {
            e.printStackTrace();
        }finally {
            try {
                if(reader!=null) reader.close();
                if(con!=null)con.disconnect();
```

```
                } catch (IOException e) {
                    e.printStackTrace();
                }
            }
        }
        private void showResult(final String result){//显示响应结果
            textView.setText(result);
        }
}
```

（5）运行项目，测试运行效果。

上述 MainActivity.java 中的 HTTP 请求代码本身没有任何问题，但运行时会出错，发生 android.os.NetworkOnMainThreadException 异常。这是因为需要在新的线程中完成比较耗时的网络访问操作。

（6）修改 MainActivity.java，创建线程来执行 HTTP 请求，代码如下。

```
package com.example.xbg.usehttpurlconnection;
import android.support.v7.app.AppCompatActivity;
...
public class MainActivity extends AppCompatActivity {
    TextViewtextView;
    @Override
    protected void onCreate(Bundle savedInstanceState) {
        super.onCreate(savedInstanceState);
        setContentView(R.layout.activity_main);
        textView= (TextView) findViewById(R.id.textView);
        new Thread(new Runnable() {//在新线程中完成HTTP请求
            @Override
            public void run() {
                doUrlGet();
            }
        }).start();
    }
    private void doUrlGet(){
        ...
    }
    private void showResult(final String result){
        runOnUiThread(new Runnable() {
            @Override
            public void run() {
                textView.setText(result);
            }
        });
    }
}
```

在上述代码中，在主活动的 onCreate()方法中调用 new Thread()方法新建一个线程来执行 HTTP 请求，处理完服务器返回结果时，调用自定义的 showResult()方法将结果显示在 TextView 控件中。

因为是在 new Thread()方法新建的线程中调用 showResult()方法，所以需要调用 runOnUiThread()返回主线程，然后才能修改 TextView 控件显示的文本。

（7）运行项目，测试运行效果。运行效果如图 7-2 所示，可以看到，HttpURLConnection 返回的是请求的 URL 的 HTML 代码。在实际应用中，开发人员可以自定义返回的数据，不一定是 HTML 代码，例如可以返回 XML 或者 JSON 字符串。

图 7-2　使用 HttpURLConnection 的运行效果

7.2.2　使用 OkHttp

HttpURLConnection 将服务器响应结果封装在 InputStream 中，需要通过编程从中读取结果。OkHttp 是 Square 公司开发的一个开源 HTTP 访问项目，相比之下使用起来非常简单。OkHttp 的主页地址为 http://square.github.io/okhttp，读者可从中了解 OkHttp 的详细信息。目前，OkHttp 的最新版本为 3.8.0。

使用 OkHttp 完成 HTTP 请求需要下面几个步骤。实例项目：源代码\07\UseOkHttp。

（1）修改项目的 app/build.gradle 文件，添加 OkHttp 库编译信息，代码如下。

```
dependencies {
    ...
    compile 'com.android.support:appcompat-v7:25.3.1'
    testCompile 'junit:junit:4.12'
    compile 'com.squareup.okhttp3:okhttp:3.8.0'
}
```

Gradle 在构建项目时，可自动下载需要的 OkHttp 相关的库文件。

（2）创建 OkHttpClient 对象，代码如下。

```
OkHttpClient okClient=new OkHttpClient();
```

（3）通过创建 Request.Builder 来创建 Request 对象，代码如下。

```
Request.Builder builder=new Request.Builder();
builder.url("https://developer.android.google.cn ");
Request request=builder.build();
```

默认 OkHttp 使用 GET 方法完成 HTTP 请求。如果要使用 POST 方法向服务器提交数据，则需要创建 RequestBody 对象来封装数据，代码如下。

```
RequestBody requestBody=new FormBody.Builder()
        .add("id","admin")
        .add("password","123")
        .build();
builder.post(requestBody);
```

（4）调用 Request 对象的 execute()方法执行请求，返回结果封装在 Response 对象中，代码如下。

```
Response response=okClient.newCall(request).execute();
```

（5）获得字符串形式的返回结果，代码如下。

```
String result=response.body().string();
```

7.3 解析 XML 格式数据

XML 已成为一种常用的数据交换格式。应用配置、应用之间交换数据或者网络数据传输，都会用到 XML 格式。http://www.w3school.com.cn/xml/index.asp 提供了一个 XML 简略教程，读者可访问学习。

在使用 HttpURLConnection、OKHttp 等执行 HTTP 请求时，使用 XML 格式来封装数据，再使用 Pull 或 DOM 等常见的 XML 解析方式，即可获得服务器返回的具体数据。

7.3.1 准备 XML 数据

在学习如何解析从服务器获得的 XML 数据之前，应先做一些准备工作，准备好服务器端的 XML 数据。

本书采用 Windows 10 自带的 IIS 作为 Web 服务器。在服务器中创建 XML 文件 getxml.xml 的具体操作步骤如下。

（1）查看是否已经启用 IIS。打开 Windows“开始”菜单，查看“Windows 管理工具”中是否有“Internet Information Services(IIS)管理器”。如果有，说明已经启用了 IIS，跳过下面的（2）~（5）步。

（2）右击任务栏左侧的 Windows 图标，在打开的快捷菜单中选择“程序和功能”命令，打开 Windows“程序和功能”窗口，如图 7-3 所示。

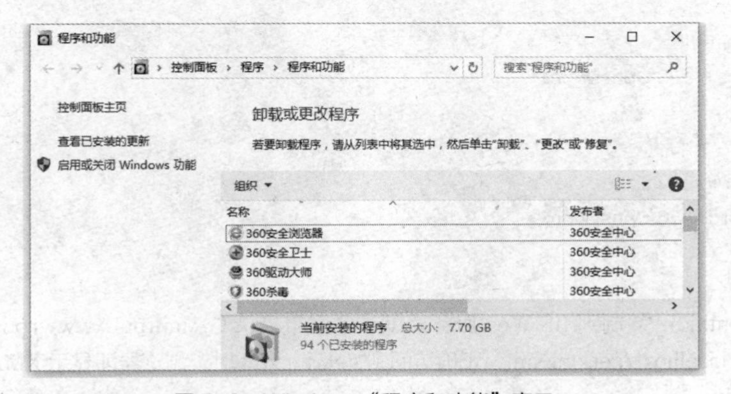

图 7-3 Windows“程序和功能”窗口

（3）单击窗口左侧的“启用或关闭 Windows 功能”选项，打开“Windows 功能”窗口，如图 7-4 所示。

图 7-4 “Windows 功能”窗口

（4）在列表框中选中 Internet Information Services 下的“Web 管理工具”和“万维网服务”选项，单击 确定 按钮，启用选中功能（系统自动安装需要的组件）。

（5）安装完成后，打开浏览器，访问 http://localhost/，如图 7-5 所示，说明 IIS 已经准确完成了安装，并启动了 Web 服务。

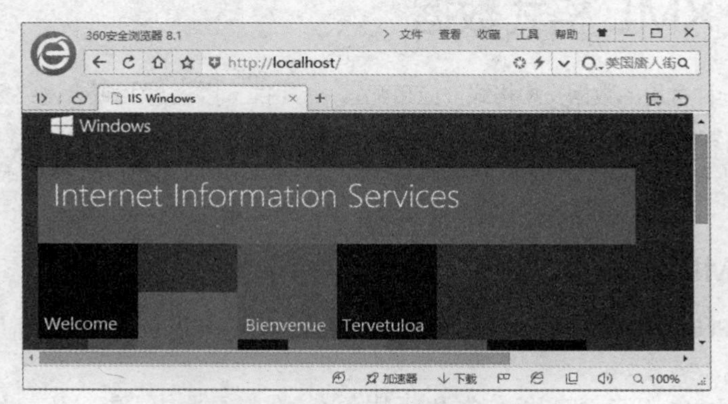

图 7-5 访问默认 Web 服务器主页

（6）使用记事本创建文件 getxml.xml（注意，保存文件时，文件类型应选择“所有文件”，然后使用.xml 扩展名，并选择 UTF-8 编码格式），内容如下。

```
<?xml version="1.0" encoding="utf-8"?>
<users>
    <user>
        <id>admin</id>
        <password>123</password>
    </user>
    <user>
        <id>jike</id>
        <password>456</password>
    </user>
</users>
```

（7）将 getxml.xml 复制到 IIS Web 服务器默认的发布目录 C:\inetpub\wwwroot 中，然后在浏览器中访问 http://localhost/getxml.xml，正确访问到 getxml.xml 时，浏览器即显示 XML 内容，如图 7-6 所示。

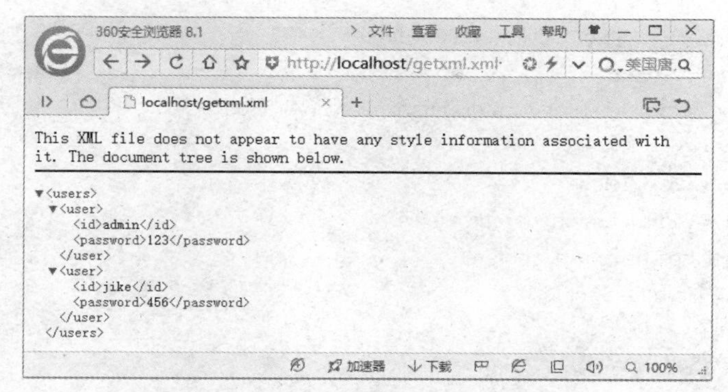

图 7-6　在浏览器中访问 XML 文件

到这里，就准备好了 XML 文件。在 Android 应用中可以通过 HTTP 请求访问该文件，然后解析出其中的数据。真机设备中需使用本机 IP 地址来访问 XML 文件，例如 http://192.168.0.104/getxml.xml。

7.3.2　DOM 解析方式

DOM 和 Pull 是 Android 中两种常用的 XML 文档解析方式，本小节先介绍 DOM。

DOM 将 XML 文档看作一个树形结构，每个标签作为一个节点。DOM 解析会遍历 XML 文档的树形结构，以获得节点和节点文本。限于篇幅，这里不再详细介绍 XML DOM，读者可访问 http://www.w3school.com.cn/xmldom/index.asp 了解详细内容。

读取与解析 XML 数据

下面通过一个实例说明如何在 Android 应用中获取并解析 XML 文档，具体操作步骤如下。实例项目：源代码\07\ParseXml。

（1）在 AndroidStudio 中创建一个新项目，将应用名称设置为 ParseXml，并为项目添加一个空活动。

（2）修改 AndroidManifest.xml 文件，申明网络访问权限，代码如下。

```
<?xml version="1.0" encoding="utf-8"?>
<manifest xmlns:android="http://schemas.android.com/apk/res/android"
    package="com.example.administrator.parsexml">
    <uses-permission android:name="android.permission.INTERNET"/>
    ...
</manifest>
```

（3）修改 app/build.gradle，添加 OkHttp 编译信息，代码如下。

```
dependencies {
    ...
    compile 'com.squareup.okhttp3:okhttp:3.8.0'
}
```

（4）修改 activity_main.xml，在主活动布局中添加控件，代码如下。

```
<?xml version="1.0" encoding="utf-8"?>
<LinearLayoutxmlns:android="http://schemas.android.com/apk/res/android"
    xmlns:tools="http://schemas.android.com/tools" android:id="@+id/activity_main"
    android:layout_width="match_parent" android:layout_height="match_parent"
    android:paddingBottom="@dimen/activity_vertical_margin"
    android:paddingLeft="@dimen/activity_horizontal_margin"
    android:paddingRight="@dimen/activity_horizontal_margin"
```

```
    android:paddingTop="@dimen/activity_vertical_margin"
    android:orientation="vertical"
    tools:context="com.example.administrator.parsexml.MainActivity">
    <Button
        android:text="获取XML文件"
        android:layout_width="match_parent"
        android:layout_height="wrap_content"
        android:id="@+id/btGetXml" />
    <TextView
        android:layout_width="match_parent"
        android:layout_height="wrap_content"
        android:text="Hello World!"
        android:id="@+id/tvXml" />
    <Button
        android:text="使用DOM解析"
        android:layout_width="match_parent"
        android:layout_height="wrap_content"
        android:id="@+id/btDomXml" />
    <TextView
        android:text="TextView"
        android:layout_width="match_parent"
        android:layout_height="wrap_content"
        android:id="@+id/tvDomResult" />
    ...
    </LinearLayout>
```

（5）修改 MainActivity.java，代码如下。

```
package com.example.administrator.parsexml;
import android.support.v7.app.AppCompatActivity;
...
public class MainActivity extends AppCompatActivity {
    TextViewtvXml;
    @Override
    protected void onCreate(Bundle savedInstanceState) {
        super.onCreate(savedInstanceState);
        setContentView(R.layout.activity_main);
        tvXml= (TextView) findViewById(R.id.tvXml);
        Button btGetXml= (Button) findViewById(R.id.btGetXml);
        btGetXml.setOnClickListener(new View.OnClickListener() {
            @Override
            public void onClick(View v) {//单击按钮时通过HTTP请求获取XML文档
                new Thread(new Runnable() {
                    @Override
                    public void run() {
                        doUrlGet();
                    }
                }).start();
```

```
            }
        });
        Button btDomXml=(Button) findViewById(R.id.btDomXml);
        btDomXml.setOnClickListener(new View.OnClickListener() {
            @Override
            public void onClick(View v) {//单击按钮时解析XML文档
                TextViewtvDomResult= (TextView) findViewById(R.id.tvDomResult);
                tvDomResult.setText(domXml());
            }
        });
        ...
    }
    private void doUrlGet(){//使用OkHttp获取XML文档
        try {
            OkHttpClientokClient=new OkHttpClient();
            Request.Builder builder=new Request.Builder();
            builder.url("http://192.168.0.104/getxml.xml");
            Request request=builder.build();
            Responseresponse=okClient.newCall(request).execute();
            showResult(response.body().string());
        } catch (Exception e) {
            e.printStackTrace();
        }
    }
    private void showResult(final String result){
        runOnUiThread(new Runnable() {//返回主线程
            @Override
            public void run() {
                tvXml.setText(result);//在TextView中显示XML文档
            }
        });
    }
    private String domXml(){//使用DOM解析XML文档
        try{
            String xmlData=tvXml.getText().toString();
            DocumentBuilderFactory factory=DocumentBuilderFactory.newInstance();
            DocumentBuilder builder=factory.newDocumentBuilder();
            InputSource data= new InputSource(new ByteArrayInputStream(xmlData.getBytes("UTF-8")));
            Document document=builder.parse(data);
            Element root=document.getDocumentElement();
            NodeList nodes=root.getElementsByTagName("user");
            String result="";
            for (int i=0;i<nodes.getLength();i++){
                Element user=(Element)nodes.item(i);
                Element id=(Element)user.getElementsByTagName("id").item(0);
                Element password=(Element) user.getElementsByTagName("password").item(0);
```

```
                result+="id="+id.getTextContent();
                result+="\npassword="+password.getTextContent();
                result+="\n";
            }
            return result;
        }catch (Exception e){
            e.printStackTrace();
            return "";
        }
    }
}
```

（6）运行项目，测试运行效果。运行效果如图 7-7 所示。

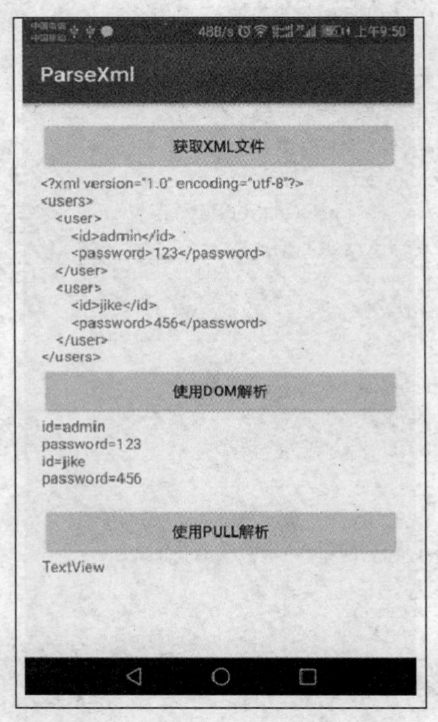

图 7-7　使用 DOM 解析 XML 文档

可以看到，在使用 DOM 解析 XML 文档时主要包括下列步骤。

（1）创建 DocumentBuilderFactory 对象。

（2）创建 DocumentBuilder 对象。

（3）将 XML 文档封装到 InputSource 对象中。

（4）使用 DocumentBuilder 对象解析 InputSource 以获得表示 XML 文档的 Document 对象。

（5）调用 Document 对象的相关方法，获取 XML 文档的各个节点及其文本。

实例中用到的 Document 对象方法如下。

- getDocumentElement：返回 XML 文档根节点。
- getElementsByTagName：返回指定名称节点的列表。
- getTextContent：返回指定名称节点的文本内容。

7.3.3 Pull解析方式

Pull解析方式将XML文档作为输入"流"来处理，依次读取每个标签，根据标签类型来处理相应数据。

1. 解析步骤

使用Pull解析XML文档的步骤如下。

（1）创建一个XmlPullParser对象作为解析器。

例如如下代码。

```
XmlPullParserFactory xmlFactory=XmlPullParserFactory.newInstance();
XmlPullParser xmlPullParser=xmlFactory.newPullParser();
```

（2）将XML文档设置为解析器的输入。

例如如下代码。

```
xmlPullParser.setInput(new StringReader(xmlData));
```

（3）获得事件类型。

Pull根据标签的类型（开始标签、结束标签）来判定事件类型。解析XML文档主要用到END_DOCUMENT（文档结束）、STAR_TAG（开始标签）和END_TAG（结束标签）3种事件类型。例如如下代码。

```
int event=xmlPullParser.getEventType();          //获得当前事件类型
event=xmlPullParser.next();                      //获得下一个事件类型
```

调用next()方法时，输入流指针前进到下一个标签位置，直到文档结束。

（4）获取当前节点数据。

如果事件类型不是文档结束，则可调用相应方法获取当前标签数据，例如如下代码。

```
String nodeName=xmlPullParser.getName();         //获得标签名称
String text=xmlPullParser.nextText()             //获得标签的文本内容
```

2. 实例解析

修改7.3.2小节中的ParseXml实例项目，添加一个按钮和一个文本视图控件用于执行解析和显示结果，即可实现使用Pull解析XML文档。MainActivity中的主要相关代码如下。实例项目：源代码\07\ParseXml。

```
package com.example.administrator.parsexml;
...
public class MainActivity extends AppCompatActivity {
    TextViewtvXml;
    @Override
    protected void onCreate(Bundle savedInstanceState) {
        ...
        Button btPullXml=(Button) findViewById(R.id.btPullXml);
        btPullXml.setOnClickListener(new View.OnClickListener() {
            @Override
            public void onClick(View v) {//单击按钮时解析XML文档
                TextViewtvPullResult= (TextView) findViewById(R.id.tvPullResult);
                tvPullResult.setText(pullXml());
            }
        });
    }
    ...
```

```
    private String pullXml(){
        try{
            String xmlData=tvXml.getText().toString();
            XmlPullParserFactory xmlFactory=XmlPullParserFactory.newInstance();
            XmlPullParserxmlPullParser=xmlFactory.newPullParser();
            xmlPullParser.setInput(new StringReader(xmlData));
            int event=xmlPullParser.getEventType(); //获得当前事件类型
            String result="",nodeName="";
            while(event!=xmlPullParser.END_DOCUMENT){
                nodeName=xmlPullParser.getName();//获得标签名称
                if(event==xmlPullParser.START_TAG){
                    if(nodeName.equals("id"))
                        result+="id="+xmlPullParser.nextText()+"\n";//获得标签文本进行处理
                    else if(nodeName.equals("password"))
                        result+="password="+xmlPullParser.nextText()+"\n";
                }
                event=xmlPullParser.next();//获得下一个事件类型
            }
            return result;
        }catch (Exception e){
            e.printStackTrace();
            return "";
        }
    }
    ...
}
```

运行效果如图 7-8 所示。

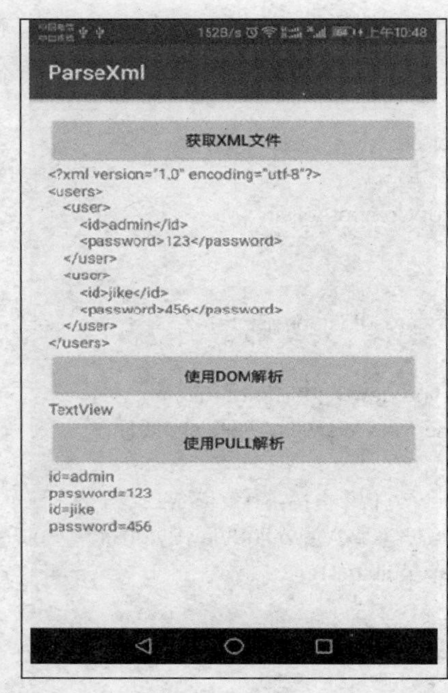

图 7-8　使用 Pull 解析 XML 的运行效果

7.4 解析 JSON 数据

读取 JSON 格式数据

JSON 主要以键值对的方式表示数据，例如如下代码。

```
{
    "jike":"极客学院",
    "users":[{"id":"admin","password":"123"},{"id":"jike","password":"456"}]
}
```

最外围的花括号表示这是一个 JSON 格式的对象数据，该对象有两个键，即 jike 和 users。jike 的值是一个字符串，users 的值是一个数组，数组有两个对象。

可以看到，JSON 与 XML 相比更简洁，可以节省网络传输时间。使用 org.json 包提供的 JSONArray、JSONObject 等类可轻松完成 JSON 数据解析。

如下代码可用于解析上述 JSON 字符串。实例项目：源代码\07\ParseJson。

```
private String JsonData(String data){
    try {
        JSONObjectjson=new JSONObject(data);
        String result="jike="+json.getString("jike")+"\n";                //获得指定键的值
        JSONArray users=json.getJSONArray("users");                       //获得指定键的数组
        for(int i=0;i<users.length();i++){
            JSONObject item=users.getJSONObject(i);                       //获得一个数组元素
            result+="user"+(i+1)+" id="+item.getString("id")+"      ";    //获取键值
            result+="password="+item.getString("password")+"\n";
        }
        return result;
    } catch (Exception e) {
        e.printStackTrace();
        return "";
    }
}
```

可以看到，在解析 JSON 数据时，首先必须对数据的结构有清晰的了解。通常，JSON 数据包含在一对花括号中，所以首先调用 JSONObject()构造方法将 JSON 字符串转换为 JSONObject 对象，然后调用 getInt()、getString()、getJSONArray()、JSONObject()等方法获取对象指定键的值。图 7-9 所示为上述实例项目 ParseJson 的运行效果。

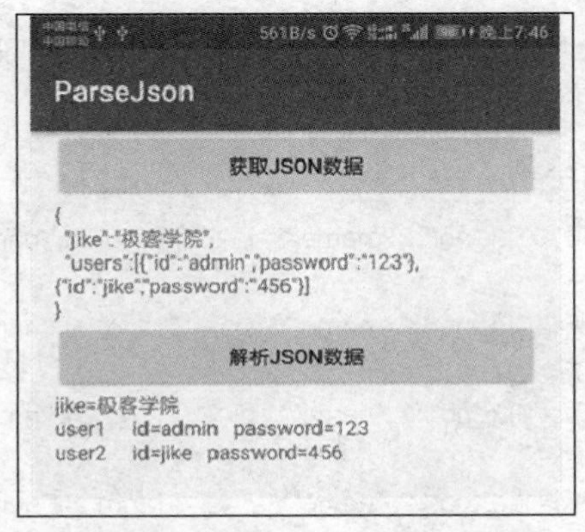

图 7-9 解析 JSON 数据的运行效果

7.5 编程实践：在线课表

本节将综合应用本章所学知识，实现 Android 在线课表功能，运行效果如图 7-10 所示。

图 7-10 在线课表运行效果

在线课表功能使用一个 Spinner 控件显示星期几，用户选择后，请求服务器端的课程安排信息，然后在 ListView 中显示当天的上课安排。

7.5.1 实现服务器端课程数据处理

在服务器端，使用 IIS 作为服务器，用一个 ASP 文件来处理客户端请求，并根据请求返回课程安排信息，操作步骤如下。

（1）参考 7.3.1 小节实例，在 IIS 中启用 ASP 功能，如图 7-11 所示。

（2）使用记事本创建一个 getclass.asp 文件，代码如下。

```
<%
    dim data(4)
    data(0)="[{""no"":1,""name"":""语文""},{""no"":2,""name"":""语文""},{""no"":3,""name"":""数学""},{""no"":5,""name"":""物理""},{""no"":6,""name"":""物理""}]"
    data(1)="[{""no"":1,""name"":""英语""},{""no"":2,""name"":""英语""},{""no"":3,""name"":""化学""},{""no"":5,""name"":""数学""},{""no"":6,""name"":""生物""}]"
    data(2)="[{""no"":1,""name"":""语文""},{""no"":2,""name"":""语文""},{""no"":4,""name"":""历史""},{""no"":5,""name"":""化学""},{""no"":6,""name"":""地理""}]"
    data(3)="[{""no"":1,""name"":""英语""},{""no"":2,""name"":""英语""},{""no"":3,""name"":""化学""},{""no"":5,""name"":""数学""},{""no"":6,""name"":""生物""}]"
```

```
    data(4)="[{""no"":1,""name"":""语文""},{""no"":2,""name"":""语文""},{""no"":4,""name"":""数学""},{""no"":5,""name"":""物理""},{""no"":6,""name"":""物理""}]"
    if Request.QueryString<>"" then
        index=Request("index")+0
        if index>=0 and index<=4 then
            response.write(data(index))
        end if
    end if
%>
```

图 7-11　启用 ASP 功能

getclass.asp 首先定义一个存放课程安排信息的 JSON 字符串的数组，然后根据客户端请求提供的星期序号返回对应课程安排信息的 JSON 字符串。

（3）将 getclass.asp 复制到 IIS 默认的 Web 发布目录 C:\inetpub\wwwroot 下，完成服务器端的设计。

7.5.2　实现 Android 在线课表

Android 在线课表的基本设计思路：使用一个下拉列表显示星期，用户在列表中选择时，将选中项的序号作为 HTTP 请求参数传递给服务器端的 getclass.asp；HTTP 请求返回包含对应星期几的课程安排信息的 JSON 字符串；解析 JSON 字符串，将课程安排信息添加到 ArrayList 中，填充 ListView，显示课程安排信息。

具体操作步骤如下。

（1）在 AndroidStudio 中创建一个新项目，将应用名称设置为 ClassSchedule，并为项目添加一个空活动。

（2）修改 AndroidManifest.xml 文件，申明网络访问权限，代码如下。

```
<?xml version="1.0" encoding="utf-8"?>
<manifest xmlns:android="http://schemas.android.com/apk/res/android"
    package="com.example.administrator.parsexml">
    <uses-permission android:name="android.permission.INTERNET"/>
    ...
</manifest>
```

（3）修改 app/build.gradle，添加 OkHttp 编译信息，代码如下。

```
dependencies {
    ...
    compile 'com.squareup.okhttp3:okhttp:3.8.0'
}
```

（4）修改 activity_main.xml，在主活动布局中添加控件，代码如下。

```
<?xml version="1.0" encoding="utf-8"?>
<LinearLayoutxmlns:android="http://schemas.android.com/apk/res/android"
    xmlns:tools="http://schemas.android.com/tools"
    android:id="@+id/activity_main"
    android:layout_width="match_parent"
    android:layout_height="match_parent"
    android:paddingBottom="@dimen/activity_vertical_margin"
    android:paddingLeft="@dimen/activity_horizontal_margin"
    android:paddingRight="@dimen/activity_horizontal_margin"
    android:paddingTop="@dimen/activity_vertical_margin"
    android:orientation="vertical"
    tools:context="com.example.administrator.classschedule.MainActivity">
    <TextView
        android:layout_width="wrap_content"
        android:layout_height="wrap_content"
        android:text="我的课表" />
    <Spinner android:id="@+id/spWeek"
        android:layout_width="match_parent"
        android:layout_height="wrap_content"/>
    <ListView android:id="@+id/lvClass"
        android:layout_width="match_parent"
        android:layout_height="wrap_content"
        android:layout_weight="1"/>
</LinearLayout>
```

（5）新建一个 Java 类 MyClass，用于封装每节课的信息，代码如下。

```
package com.example.administrator.classschedule;
/**
 * MyClass类用于封装每节课的信息，包括节次（no）、上课时间（time）和课程名称（name）
 */
public class MyClass {
    private String no,time,name;
    public MyClass(String no, String time, String name) {
        this.no = no;
        this.time = time;
        this.name = name;
    }
    public String getNo() {return no;    }
    public void setNo(String no) { this.no = no;    }
```

```
    public String getTime() { return time;     }
    public void setTime(String time) { this.time = time;     }
    public String getName() { return name;     }
    public void setName(String name) { this.name = name;   }
}
```

（6）新建一个布局文件 classitem.xml，定义显示课程安排的 ListView 的每个列表项的布局，代码如下。

```
<?xml version="1.0" encoding="utf-8"?>
<LinearLayoutxmlns:android="http://schemas.android.com/apk/res/android"
    xmlns:tools="http://schemas.android.com/tools"
    android:orientation="horizontal" android:layout_width="match_parent"
    android:layout_height="match_parent">
    <TextView
        android:text="TextView"
        android:layout_width="100dp"
        android:layout_height="wrap_content"
        android:gravity="center_horizontal"
        android:id="@+id/tvNo"
        android:textStyle="bold" />
    <TextView
        android:text="TextView"
        android:layout_width="100dp"
        android:layout_height="wrap_content"
        android:gravity="center_horizontal"
        android:id="@+id/tvTime"/>
    <TextView
        android:text="TextView"
        android:layout_width="wrap_content"
        android:layout_height="wrap_content"
        android:gravity="center_horizontal"
        android:id="@+id/tvName"
        android:layout_weight="1" />
</LinearLayout>
```

（7）新建一个 Java 类 ClassAdapter，实现显示课程安排的 ListView 的适配器，代码如下。

```
package com.example.administrator.classschedule;
import android.content.Context;
…
public class ClassAdapter extends ArrayAdapter<MyClass> {
    public int resId;
    publicClassAdapter(Context context, int resource, List<MyClass> objects) {
        super(context, resource, objects);
        resId=resource;
    }
    public View getView(int position, View convertView, ViewGroup parent) {
```

```
        MyClass one=getItem(position);//获得当前列表项MyClass对象
        View view= LayoutInflater.from(getContext()).inflate(resId,parent,false);
        TextView tvNo=(TextView)view.findViewById(R.id.tvNo);
        tvNo.setText(one.getNo());
        TextView tvTime=(TextView)view.findViewById(R.id.tvTime);
        tvTime.setText(one.getTime());
        TextView tvName=(TextView)view.findViewById(R.id.tvName);
        tvName.setText(one.getName());
        return view;
    }
}
```

（8）修改 MainActivity.java，代码如下。

```
package com.example.administrator.classschedule;
import android.support.v7.app.AppCompatActivity;
...
public class MainActivity extends AppCompatActivity {
    private static String[] week={"星期一","星期二","星期三","星期四","星期五"};
    private static String[] nos={"第1节","第2节","第3节","第4节","第5节","第6节","第7节"};
    private static String[] times={"08:00-08:50","09:00-09:50","10:00-10:50",
                                        "11:00-11:50","14:00-14:50","15:00-15:50","16:00-16:50"};
    privateList<MyClass> schedule=null;     //课程安排列表
    @Override
    protected void onCreate(Bundle savedInstanceState) {
        super.onCreate(savedInstanceState);
        setContentView(R.layout.activity_main);
        //初始化星期下拉列表Spinner
        ArrayAdapter<String> adapter =
                new ArrayAdapter<String>(this, android.R.layout.simple_list_item_1, week);
        Spinner spWeek=(Spinner)findViewById(R.id.spWeek);
        spWeek.setAdapter(adapter);
        spWeek.setOnItemSelectedListener(new AdapterView.OnItemSelectedListener() {
            @Override
            public void onItemSelected(AdapterView<?> parent, View view, final int position, long id) {
                new Thread(new Runnable() {//在新线程中完成HTTP请求
                    @Override
                    public void run() {
                        showClass(doUrlGet(position));//选择星期几时，显示课程安排
                    }
                }).start();
            }
            @Override
            public void onNothingSelected(AdapterView<?> parent) {}
        });
    }
```

```
    private String doUrlGet(int index){//使用OkHttp获取服务器端课程安排信息的JSON数据
        try {
            OkHttpClientokClient=new OkHttpClient();
            Request.Builder builder=new Request.Builder();
            builder.url("http://192.168.0.104/getclass.asp?index="+index);
            Request request=builder.build();
            Responseresponse=okClient.newCall(request).execute();
            return response.body().string();
        } catch (Exception e) {
            e.printStackTrace();
            return "";
        }
    }
    private void showClass(final String data){
        runOnUiThread(new Runnable() {//返回主线程
            @Override
            public void run() {
                schedule=new ArrayList<MyClass>();
                //初始化课程信息安排，每天7节课，节次、时间固定，课程名称初始化为空
                for(int i=0;i<7;i++)
                    schedule.add(new MyClass(nos[i],times[i],""));
                try {//解析JSON字符串中的课程安排信息，按节次将课程名称补充完整
                    JSONArrayjsonArray = new JSONArray(data);
                    for (int i = 0; i <jsonArray.length(); i++) {
                        String cname;
                        JSONObject item = jsonArray.getJSONObject(i);
                        int n=item.getInt("no");
                        cname=item.getString("name");            //获得课程名称
                        schedule.get(n-1).setName(cname);        //修改列表中的名称
                    }
                    ClassAdapter adapter=new ClassAdapter(MainActivity.this,
                            R.layout.classitem,schedule);//创建适配器
                    ListView listView=(ListView)findViewById(R.id.lvClass);
                    listView.setAdapter(adapter);//用适配器填充课程安排ListView
                }catch(Exception e){
                    e.printStackTrace();
                }
            }
        });
    }
}
```

（9）运行程序，测试运行效果。

7.6 小结

网络访问不外乎浏览网页和请求服务器端数据，利用 WebView 控件可轻松实现具有浏览器功

能的 Android 应用来浏览网页，请求服务器端数据可使用 Android 内置的 HttpURLConnect 或第三方组件（如 OkHttp）。HttpURLConnect 将从服务器返回的数据封装在 InputStream 对象中，需要编程解析才能获得最终数据。OkHttp 封装了这些底层操作，调用一个方法即可获得服务器端返回的字符串。

为了更有效率地处理服务器端返回的数据，通常采用 XML 或 JSON 格式来表示数据。在 Android 客户端，可使用内置的 DOM 或 Pull 来解析 XML。要解析 JSON 数据，可使用 org.json 包提供的 JSONArray、JSONObject 等类。

7.7 习题

1. 简述使用 HttpURLConnection 完成 HTTP 请求的基本步骤。
2. 简述使用 OkHttp 完成 HTTP 请求的基本步骤。
3. 简述使用 DOM 解析 XML 文件的基本步骤。

PART08

第8章

线程和服务

重点知识：

- 线程
- 服务

■ 早期的 iOS 和 Windows Phone 系统不支持后台功能，这意味着用户在打电话时不能够听音乐或者在后台挂着 QQ、微信等。Android 吸取了 Symbian 系统的优点，使用“服务”来实现后台功能，支持后台功能。所以 Java 程序员应熟悉多线程编程，在子线程中完成比较耗时的工作。本章将介绍如何使用线程和服务。

8.1 多线程

在运行一个 Android 应用时，系统会为其创建一个独立主线程。在程序执行一些比较耗时的操作（如打开网页）时，应用界面将无法响应用户操作。如果将耗时操作放到子线程中去执行，子线程与主线程异步同时运行，那么当子线程去执行耗时操作时，用户可在界面中执行其他操作。

线程的概念

8.1.1 线程的基本用法

1. 基本用法

（1）Android 的多线程比较简单，第 7 章中使用了最简单的创建线程的方法，即使用匿名类，代码如下。

```
new Thread(new Runnable() {
    @Override
    public void run() {
        //在此编写线程功能代码
    }
}).start();
```

new Thread()方法创建了一个线程对象，然后调用 start()方法启动线程。new Runnable() {}创建了一个匿名类来实现 Runnable 接口，需要在其 run()方法中编写实现线程功能的代码。

（2）另外，也可创建一个类来实现 Runnable 接口，代码如下。

```
classMyThread implements Runnable{
    @Override
    public void run() {
        //在此编写线程功能代码
    }
}
```

然后按照下面的方式来启动线程。

```
new Thread(new MyThread()).start();
```

（3）还可定义一个类继承内置的 Thread 类来实现线程功能，代码如下。

```
classMyThread extends Thread{
    @Override
    public void run() {
        //在此编写线程功能代码
    }
}
```

然后按照下面的方式来启动线程。

```
newMyThread().start();
```

2. 实例详解

下面通过具体的实例说明如何使用多线程。实例项目：源代码\08\ThreadTest。

（1）创建一个 ThreadTest 项目，然后修改 activity_main.xml 定义布局，用 3 个按钮测试前面介绍的 3 种方法来创建线程。activity_main.xml 代码如下。

```
<?xml version="1.0" encoding="utf-8"?>
<LinearLayoutxmlns:android="http://schemas.android.com/apk/res/android"
```

```
...
    tools:context="com.example.xbg.threadtest.MainActivity">
    <Button
        android:text="匿名线程类"
        android:layout_width="match_parent"
        android:layout_height="wrap_content"
        android:id="@+id/btSimple" />
    <Button
        android:text="线程子类"
        android:layout_width="match_parent"
        android:layout_height="wrap_content"
        android:id="@+id/btThread" />
    <Button
        android:text="实现Runnable接口"
        android:layout_width="match_parent"
        android:layout_height="wrap_content"
        android:id="@+id/btRunnable" />
</LinearLayout>
```

（2）修改 MainActivity.java，在 3 个按钮的单击事件监听器中用不同的方法创建线程，代码如下。

```
package com.example.xbg.threadtest;
import android.support.v7.app.AppCompatActivity;
...
public class MainActivity extends AppCompatActivity {
    @Override
    protected void onCreate(Bundle savedInstanceState) {
        super.onCreate(savedInstanceState);
        setContentView(R.layout.activity_main);
        Button btSimple=(Button)findViewById(R.id.btSimple);
        btSimple.setOnClickListener(new View.OnClickListener() {
            @Override
            public void onClick(View v) {
                new Thread(new Runnable() {//使用匿名线程类
                    @Override
                    public void run() {
                        for(int i=1;i<4;i++){
                            Log.e("btSimple.click()","匿名使用线程！");
                            try {
                                Thread.sleep(1000);//线程暂停1s
                            } catch (InterruptedException e) {
                                e.printStackTrace();
                            }
                        }
                    }
                }).start();
            }
```

```
        });
        Button btThread=(Button)findViewById(R.id.btThread);
        btThread.setOnClickListener(new View.OnClickListener() {
            @Override
            public void onClick(View v) {
                new MyThread1().start();//使用自定义Thread子类
            }
        });
        Button btRunnable=(Button)findViewById(R.id.btRunnable);
        btRunnable.setOnClickListener(new View.OnClickListener() {
            @Override
            public void onClick(View v) {
                new Thread(new MyThread2()).start();//使用实现Runnable接口的类创建线程
            }
        });
    }
    class MyThread1 extends Thread{//自定义Thread子类
        @Override
        public void run() {
            for(int i=1;i<4;i++){
                Log.e("btThread.click()","使用自定义Thread子类！" );
                try {
                    Thread.sleep(1000);//线程暂停1s
                } catch (InterruptedException e) {
                    e.printStackTrace();
                }
            }
        }
    }
    classMyThread2 implements Runnable{
        @Override
        public void run() {
            for(int i=1;i<4;i++){
                Log.e("btRunnable.click()","使用自定义类实现Runnable接口" );
                try {
                    Thread.sleep(1000);//线程暂停1s
                } catch (InterruptedException e) {
                    e.printStackTrace();
                }
            }
        }
    }
}
```

项目运行界面如图 8-1 所示，程序中的 3 种方法创建的线程完成了相同的任务，首先调用 Log.e() 方法输出调试信息，然后调用 Thread.sleep()方法使线程暂停 1s。

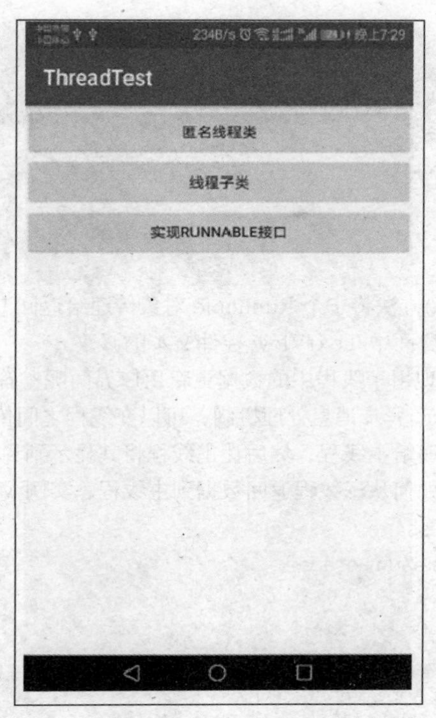

图 8-1　使用不同方法创建线程的运行界面

依次单击 3 个按钮，在 AndroidStudio 的 Run 窗口中可看到 3 个子线程输出的调试信息，如图 8-2 所示。从输出结果可以看到，首先执行的是用匿名类方式创建的线程，系统并没有等到该线程全部执行完再去执行其他线程，而是在线程休眠期间执行了其他线程。

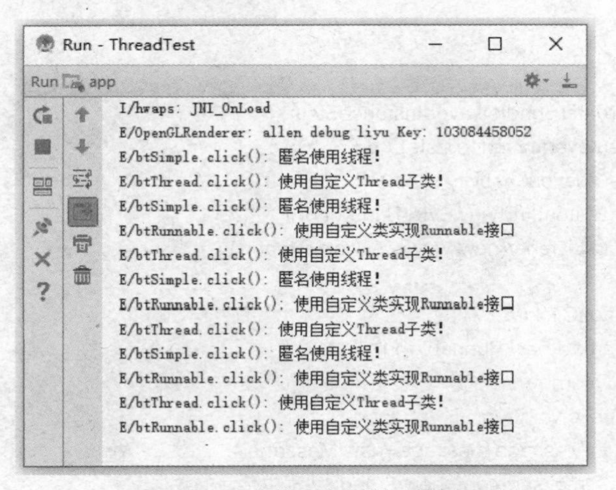

图 8-2　3 个线程输出的调试信息

8.1.2　如何在使用多线程时更新 UI

Android 中不允许在主线程之外的子线程中修改应用界面，例如试图在子线程中将处理结果显示在 TextView 中，这样做会导致程序出现异常。

回顾以下第 7 章中的实例代码。

```
private void showResult(final String result){
    runOnUiThread(new Runnable() {
        @Override
        public void run() {
            textView.setText(result);
        }
    });
}
```

代码中，runOnUiThread()方法将一个 Runnable 对象传递给返回 UI 线程（也就是主线程），并通过其中的 run()方法完成对 UI 界面中 TextView 控件文本的修改。

还有没有其他方法可以实现用子线程中的数据更新用户界面呢？答案是肯定的。多线程是异步执行的，所以 Android 提供了一套异步消息处理机制，可以在线程之间传递消息。利用这一特点，可以将子线程中的数据通过消息传递给主线程，然后让主线程将其显示到界面中。

下面先通过一个实例介绍如何从子线程返回数据到主线程，实例 MainActivity 代码如下。实例项目：源代码\08\UseThreadMessage。

```
package com.example.xbg.usethreadmessage;
import android.os.Handler;
import android.os.Message;
…
public class MainActivity extends AppCompatActivity {
    private Handler handler=new Handler(){
        @Override
        public void handleMessage(Message msg) {
            TextViewtextView=(TextView)findViewById(R.id.tvMsg);
            textView.setText(msg.obj.toString());
        }
    };
    @Override
    protected void onCreate(Bundle savedInstanceState) {
        super.onCreate(savedInstanceState);
        setContentView(R.layout.activity_main);
        Button button=(Button)findViewById(R.id.button);
        button.setOnClickListener(new View.OnClickListener() {
            @Override
            public void onClick(View v) {
                new Thread(new Runnable() {
                    @Override
                    public void run() {
                        Message message=new Message();
                        message.obj=new String("线程中传回的数据");
                        handler.sendMessage(message);
                    }
                }).start();
            }
        });
    }
}
```

程序运行效果如图 8-3 所示。

主活动 MainActivity 中定义了一个 Handler 对象，其 handleMessage()方法在对象接收到消息时调用。单击按钮时启动一个子线程。线程中创建了一个 Message 对象，并在其 obj 字段中保存了一个字符串对象，然后调用 Handler 对象的 sendMessage()方法发送消息。因为 Handler 对象是在主线程中创建的，所以它接收到消息时也在主线程中执行 handleMessage()方法，因而可以在 handleMessage()方法中更新界面。

Android 中线程之间的消息传递主要由 Message（消息）、Handler（消息处理器）、MessageQueue（消息队列）和 Looper（消息循环）来完成。

- Message 用于封装消息，它的 arg1、arg2 和 what 字段用于存放 int 类型数据，obj 字段用于存放任意类型的对象。
- Handler 主要用于发送和处理消息，通常在子线程中调用 sendMessage()方法发送消息，在主线程中执行 handleMessage()方法处理消息。消息的发送和处理是异步执行的，不能期望消息发送之后 Handler 立即处理消息。
- 通过 Handler 发送的消息都保存在 MessageQueue 中，等待被处理。
- Looper 主要完成消息派遣任务。Looper 维持一个无限循环，不停地检查消息队列中是否存在消息。当 Looper 发现消息队列中有消息时，就将队列最前面的消息取出，传递给 Handler。

图 8-3　从线程返回数据的运行效果

8.1.3　使用 AsyncTask

AsyncTask 的使用方法

AsyncTask 是 Android 为了简化使用线程数据更新 UI 而提供的一个抽象类。使用 AsyncTask，不需要了解线程和异步消息处理机制即可完成异步任务的执行。

AsyncTask 是一个抽象类，在使用时需要创建一个类来继承它，例如如下代码。

```
private class  MyAsyncTask extends AsyncTask<int[],String,String>{
        @Override
        protected void onPreExecute() {
            //异步任务开始执行之前执行的代码
        }
        @Override
        protected void onPostExecute(String s) {
            //异步任务执行结束之后执行的代码
        }
        @Override
        protected void onProgressUpdate(String... values) {
            //异步任务执行过程中执行的代码
        }
        @Override
        protected String doInBackground(int[]... params) {
            //异步任务代码
        }
    }
```

在继承 AsyncTask 时，首先需要指定 3 个泛型参数，其作用分别如下。

- 第 1 个泛型参数：指定 doInBackground()方法参数 params 的数据类型。参数 params 也称传入参数，保存调用 AsyncTask 子类构造函数时传入的参数。
- 第 2 个泛型参数：指定 onProgressUpdate()方法参数 values 的数据类型。参数 values 保存异步任务执行过程中传递回来的数据。
- 第 3 个泛型参数：指定 onPostExecute()方法参数和 doInBackground()方法返回值的数据类型。

此外，还需重写下述几个方法以完成相应任务。

- onPreExecute()方法：在异步任务开始执行之前被调用，并在主线程中运行。
- onPostExecute()方法：在异步任务执行结束之后被调用，并在主线程中运行。
- onProgressUpdate()方法：异步任务代码中可调用 publishProgress()方法向主线程返回数据，onProgressUpdate()方法参数接收返回的数据。onProgressUpdate()方法也在主线程中执行。
- doInBackground()方法：异步任务代码在子线程中执行。onPostExecute()方法参数接收 doInBackground()方法的返回值。

下述实例实现了从子线程返回数据到主线程。实例项目：源代码\08\UseAsyncTask。

（1）创建项目 UseAsyncTask，然后修改 activity_main.xml，代码如下。

```
<?xml version="1.0" encoding="utf-8"?>
<LinearLayoutxmlns:android="http://schemas.android.com/apk/res/android"
    ...
    tools:context="com.example.xbg.useasynctask.MainActivity">
    <Button
        android:text="启动异步任务"
        android:layout_width="match_parent"
        android:layout_height="wrap_content"
        android:id="@+id/button" />
    <TextView
        android:layout_width="wrap_content"
```

```
        android:layout_height="wrap_content"
        android:text=""
        android:id="@+id/tvStart" />
    <TextView
        android:text=""
        android:layout_width="match_parent"
        android:layout_height="wrap_content"
        android:id="@+id/tvProgress" />
    <TextView
        android:text=""
        android:layout_width="match_parent"
        android:layout_height="wrap_content"
        android:id="@+id/tvResult" />
</LinearLayout>
```

按钮 button 用于启动异步任务，3 个文本视图分别用于显示异步任务开始执行、执行进度和执行结果消息。

（2）修改 MainActivity.java，定义一个内部类继承 AsyncTask 来实现异步任务，代码如下。

```
package com.example.xbg.useasynctask;
import android.os.AsyncTask;
...
public class MainActivity extends AppCompatActivity {
    Private TextView tvStart,tvProgress,tvResult;
    @Override
    protected void onCreate(Bundle savedInstanceState) {
        super.onCreate(savedInstanceState);
        setContentView(R.layout.activity_main);
        tvStart=(TextView)findViewById(R.id.tvStart);
        tvProgress=(TextView)findViewById(R.id.tvProgress);
        tvResult=(TextView)findViewById(R.id.tvResult);
        Button button= (Button) findViewById(R.id.button);
        button.setOnClickListener(new View.OnClickListener() {
            @Override
            public void onClick(View v) {
                tvStart.setText("");
                tvResult.setText("");
                tvProgress.setText("");
                int[] data={1,2,3,4,5};
                new MyAsyncTask().execute(data);//启动异步任务
            }
        });
    }
    private class   MyAsyncTask extends AsyncTask<int[],String,String>{
        @Override
        protected void onPreExecute() {tvStart.setText("后台任务开始执行....");  }
        @Override
```

```
        protected void onPostExecute(String s) { tvResult.setText(s); }
        @Override
        protected void onProgressUpdate(String... values) { tvProgress.setText(values[0]); }
        @Override
        protected String doInBackground(int[]... params) {
            int s=0;
            for(int i=0;i<params[0].length;i++){
                try {
                    publishProgress("当前进度："+(int)((i+1)*100/params[0].length)+"%");
                    Thread.sleep(1000);
                } catch (InterruptedException e) {
                    e.printStackTrace();
                }
                s+=params[0][i];
            }
            return "后台计算结果："+s;
        }
    }
}
```

图 8-4 所示为上述实例程序的执行效果。

图 8-4　执行异步任务执行效果

8.2　服务

通常，一个应用通过 UI 与用户进行交互。一些特殊的应用，例如与 Web 服务器的数据传输、下载文件、与服务器保持推送连接等，并不需要用户界面，这种应用就可使用服务来实现。

8.2.1 使用服务

使用 Service

在 AndroidStudio 中，右击 MainActivity.java 所在的文件夹，如 com.example.xbg.useservice，然后在弹出的快捷菜单中选择“New\Service\Service”命令，打开新建 Android 组件对话框，如图 8-5 所示。

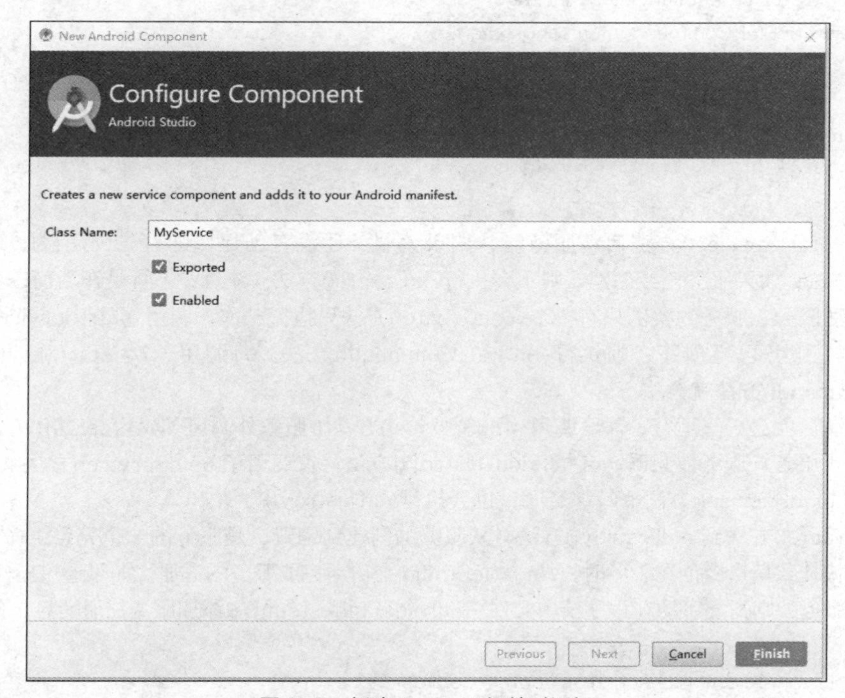

图 8-5　新建 Android 组件对话框

在 Class Name 文本框中输入服务名称，如“MyService”，单击 Finish 按钮完成新建 Android 组件操作。勾选 Exported 复选框，表示允许当前应用之外的应用访问该服务；勾选 Enabled 复选框，表示启用该服务。

AndroidStudio 创建的服务类代码如下。

```
package com.example.xbg.useservice;
import android.app.Service;
import android.os.IBinder;
import android.content.Intent;
public class MyService extends Service {
    publicMyService() {
    }
    @Override
    publicIBinderonBind(Intent intent) {
        // TODO: Return the communication channel to the service.
        throw new UnsupportedOperationException("Not yet implemented");
    }
}
```

新建的类继承了 Service 类，说明这是一个服务。onBind()方法将在使用绑定服务时用到。

在实现服务的具体功能时，通常还需要重写 Service 的下列方法。

```
public void onCreate() {
    super.onCreate();
}
@Override
public int onStartCommand(Intent intent, int flags, int startId) {
    return super.onStartCommand(intent, flags, startId);
}
@Override
public void onDestroy() {
    super.onDestroy();
}
```

在调用 startService()方法启动服务时，如果该服务还没有创建，则首先创建该服务，并执行 onCreate()方法；如果服务已经创建，则不会执行 onCreate()方法。注意，不管是在当前应用还是在其他应用中启动服务，服务的实例只有一个，onCreate()方法只执行一次。调用 startService()方法启动服务时，如果服务已经创建，则执行 onStartCommand()方法。每调用一次 startService()方法，onStartCommand()方法就会执行一次。

服务启动后就会一直运行，需要调用 stopService()方法（服务外调用）或 stopSelf()方法（服务内调用）来停止服务。服务停止时，会执行 onDestroy()方法。在调用了 bindService()方法绑定了服务，然后调用 unbindService()方法解除绑定时，也会执行 onDestroy()方法。

从 onCreate()方法到 onDestroy()方法，从服务的创建到销毁，是服务的一个完整生命周期。

在通过快捷菜单命令创建服务时，AndroidStudio 会自动完成服务的注册。如果通过新建 Java 类的方式来创建服务，则需要手动在程序清单文件 AndroidManifest.xml 中添加服务注册消息，代码如下。

```
<?xml version="1.0" encoding="utf-8"?>
<manifest xmlns:android="http://schemas.android.com/apk/res/android"
    package="com.example.xbg.useservice">
    <application...>
        ...
        <service
            android:name=".MyService"
            android:enabled="true"
            android:exported="true"></service>
    </application>
</manifest>
```

实现了服务类后，就可通过调用 startService()方法启动服务，代码如下。

```
startService(new Intent(MainActivity.this,MyService.class));
```

停止服务时，会调用 stopService()方法，代码如下。

```
stopService(new Intent(MainActivity.this,MyService.class));
```

注意，在启动和停止服务时，虽然使用的是新建的 Intent 对象，但访问的是同一个服务，因为服务实例始终只有一个。

下述实例说明了如何使用服务。实例项目：源代码\08\UseService。

（1）新建一个项目 UseService，然后修改 activity_main.xml，在布局中添加两个按钮，一个用于启动服务，一个用于停止服务，代码如下。

```
<?xml version="1.0" encoding="utf-8"?>
```

```
<LinearLayoutxmlns:android="http://schemas.android.com/apk/res/android"
...
    tools:context="com.example.xbg.useservice.MainActivity">
    <Button
        android:text="启动服务"
        android:layout_width="match_parent"
        android:layout_height="wrap_content"
        android:id="@+id/btStart" />
    <Button
        android:text="停止服务"
        android:layout_width="match_parent"
        android:layout_height="wrap_content"
        android:id="@+id/btStop" />
</LinearLayout>
```

（2）再新建一个 Service 子类，代码如下。

```
package com.example.xbg.useservice;
import android.app.Service;
import android.content.Intent;
import android.os.IBinder;
import android.util.Log;
public class MyService extends Service {
    private int count=0;
    publicMyService() {
    }
    @Override
    publicIBinderonBind(Intent intent) {
        // TODO: Return the communication channel to the service.
        throw new UnsupportedOperationException("Not yet implemented");
    }
    @Override
    public void onCreate() {
        super.onCreate();
        Log.e("MyService", "服务创建完成......MyService.count="+count);
    }
    @Override
    public int onStartCommand(Intent intent, int flags, int startId) {
        count++;
        Log.e("MyService", "服务运行中......MyService.count="+count);
        return super.onStartCommand(intent, flags, startId);
    }
    @Override
    public void onDestroy() {
        super.onDestroy();
        Log.e("MyService", "服务已停止");
    }
}
```

上述代码为 MyService 定义了一个字段 count 来计算服务的启动次数，并在各个方法中使用 Log.e()方法输出测试信息。

（3）修改 MainActivity.java，实现服务的启动和停止，代码如下。

```
package com.example.xbg.useservice;
import android.content.Intent;
import android.support.v7.app.AppCompatActivity;
import android.os.Bundle;
import android.view.View;
public class MainActivity extends AppCompatActivity {
    private Intent toMyService;
    @Override
    protected void onCreate(Bundle savedInstanceState) {
        super.onCreate(savedInstanceState);
        setContentView(R.layout.activity_main);
        toMyService=new Intent(MainActivity.this,MyService.class);
        findViewById(R.id.btStart).setOnClickListener(new View.OnClickListener() {
            @Override
            public void onClick(View v) {//启动服务
                startService(new Intent(MainActivity.this,MyService.class));
            }
        });
        findViewById(R.id.btStop).setOnClickListener(new View.OnClickListener() {
            @Override
            public void onClick(View v) {//停止服务
                stopService(new Intent(MainActivity.this,MyService.class));
            }
        });
    }
}
```

程序运行时，可多次单击按钮来启动或停止服务。图 8-6 所示为输出的测试信息。可以看到，首次启动服务时才会执行 onCreate()方法，输出信息中 count 变量的值为 0。每次调用 startService()方法来启动服务都会执行 onStartCommand()方法，count 变量的值增加 1。

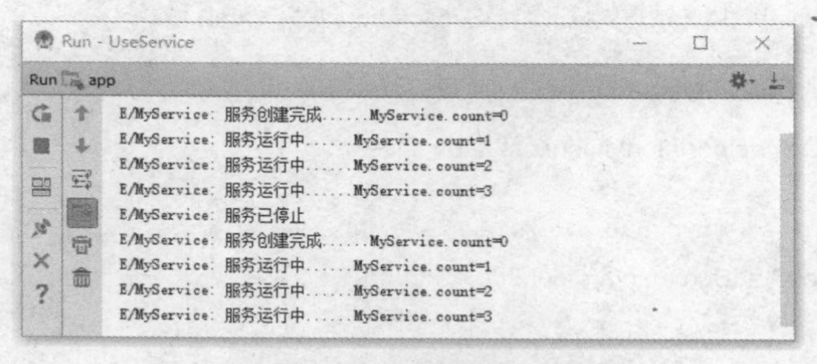

图 8-6　服务方法调用时输出的测试信息

8.2.2　使用绑定服务

8.2.1 小节中介绍的服务使用方法可以称为服务的普通用法，在这种方法下，活动对服务控制只有启动和停止操作，服务中的代码如何执行与活动没有任何关系。

Android 提供了一种可以让活动和服务进行交互的方法——绑定服务。使用绑定服务，活动可以主动启动服务操作，并从服务返回数据。

绑定 Service

在实现服务类时，onBind()方法返回一个 IBinder 对象。该对象通常是一个自定义的 Binder 子类的实例对象。通过 IBinder 对象，可以在活动中让任务完成指定操作。

要使用绑定类，首先需要实现服务类，并通过 onBind()方法返回绑定对象。例如如下代码。实例项目：源代码\08\UseBindService。

```
package com.example.xbg.usebindservice;
import android.app.Service;
...
public class MyService extends Service {
    publicMyService() {}
    @Override
    publicIBinderonBind(Intent intent) {
        return new MyBinder();//返回自定义绑定对象
    }
    class MyBinder extends Binder{//自定义绑定类
        private int result=0;
        public void startDoSomething(int[] data){
            Log.e("MyService","MyBinder.startDoSomething()方法执行...");
            for(int i=0;i<data.length;i++) result+=data[i];
        }
        public int getResult(){return   result;}
    }
}
```

该例中，内部类 MyBinder 继承了 Binder 类；startDoSomething()方法对传入的数组求和，并输出调试信息；getResult()返回求和结果。

在活动中，使用 BindService ()方法来绑定服务，该方法的第 1 个参数为启动服务的 Intent 对象，第 2 个参数为实现了 ServiceConnection 接口的类对象。ServiceConnection 接口的 onServiceConnected()方法实现在完成服务绑定后执行的操作，onServiceDisconnected()方法实现解除绑定后执行的操作。例如如下代码。

```
package com.example.xbg.usebindservice;
import android.content.ComponentName;
...
public class MainActivity extends AppCompatActivity implements View.OnClickListener   {
    classMyServiceConnection implements ServiceConnection{
        @Override
        public void onServiceConnected(ComponentName name, IBinder service) {
            Log.e("MainActivity","服务绑定完成");
            MyService.MyBindermyBinder= (MyService.MyBinder) service;
            myBinder.startDoSomething(new int[]{1,2,3,4,5});
            Log.e("MainActivity","服务返回数据："+ myBinder.getResult());
        }
        @Override
        public void onServiceDisconnected(ComponentName name) {}
```

```
    }
    privateMyServiceConnectionmyConnection=new MyServiceConnection();
    @Override
    protected void onCreate(Bundle savedInstanceState) {
        super.onCreate(savedInstanceState);
        setContentView(R.layout.activity_main);
        findViewById(R.id.btBindService).setOnClickListener(this);
        findViewById(R.id.btUnBindService).setOnClickListener(this);
    }
    @Override
    public void onClick(View v) {
        switch(v.getId()){
            case R.id.btBindService://执行绑定服务操作
                Intent intent=new Intent(this,MyService.class);
                bindService(intent,myConnection,BIND_AUTO_CREATE);
                break;
            case R.id.btUnBindService://执行解除绑定操作
                unbindService(myConnection);
                break;
        }
    }
}
```

上述实例程序运行效果如图 8-7 所示。单击按钮执行绑定服务操作后，在 Android Studio 的 Run 窗口中可看到输出的调试信息。可以看到，活动中通过绑定对象让服务完成了指定操作，并获得返回的操作结果。

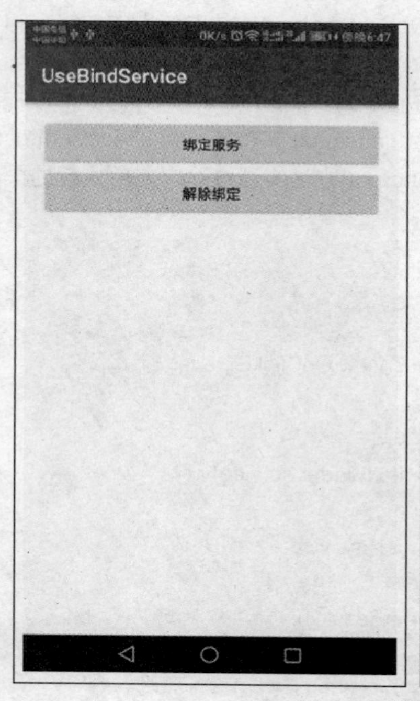

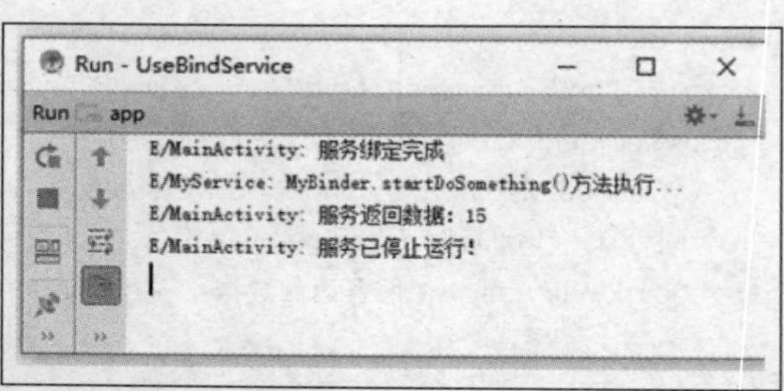

图 8-7　使用绑定服务的程序运行效果

8.3 编程实践：多线程断点续传下载

项目演示和基本原理讲解

多线程下载的实现

断点续传的实现

下载应用的完善

本节将综合应用本章所学知识，实现多线程断点续传下载，如图 8-8 所示。

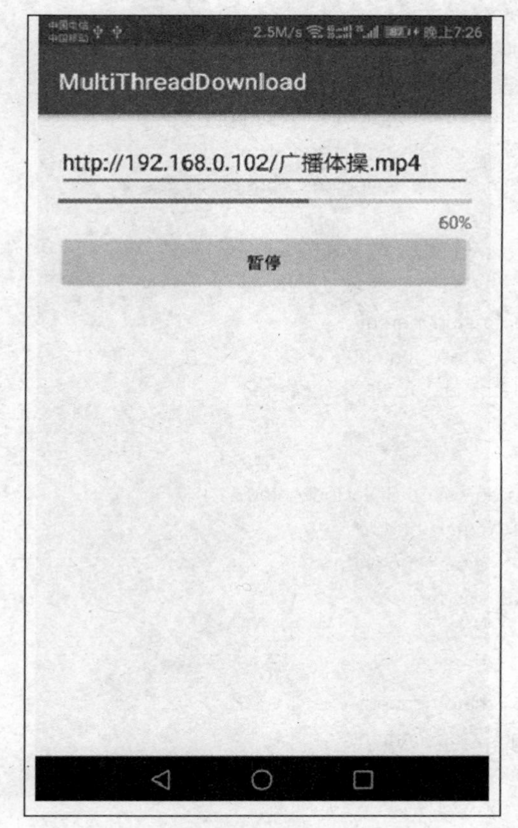

图 8-8　多线程断点续传下载

本例中，文件下载使用第 7 章中介绍的 HttpURLConnection 来完成。在 HTTP 协议头的 Range 参数中可指定获取文件的开始位置和结束位置，从而实现断点续传。例如，Range:bytes=100-500 表示服务器将从第 100 个字节开始到第 500 个字节结束的文件内容返回给客户端。使用 HttpURLConnection 对象时，可按下面的方式设置文件返回范围。

```
conn.setRequestProperty("Range","bytes=100-500");
```

利用这一特点，在下载文件时就可将文件分成多个部分，在多个线程中进行下载，提高下载效率。

分段下载时，为了在下载完成后获得一个完整的文件，可使用 RandomAccessFile。RandomAccessFile 按指定位置读写文件，所以可在多个线程中将从 HttpURLConnection 返回的

InputStream 中获得内容并分别写入文件的指定位置。

断点续传功能需要记住各个部分的下载位置。在重新开始时，断点位置作为 Range 参数的开始位置。该功能实现的具体操作步骤如下。

（1）参考 7.3.1 小节，在本地计算机中启动 IIS 服务器，并准备用于测试下载的文件。

（2）在 AndroidStudio 中创建一个新项目，将应用名称设置为 Multithreaddownload，并为项目添加一个空活动。

（3）修改 activity_main.xml，为主活动布局添加控件，代码如下。

```
<?xml version="1.0" encoding="utf-8"?>
<LinearLayoutxmlns:android="http://schemas.android.com/apk/res/android"
    xmlns:tools="http://schemas.android.com/tools"
    android:id="@+id/activity_main"
    android:layout_width="match_parent"
    android:layout_height="match_parent"
    android:paddingBottom="@dimen/activity_vertical_margin"
    android:paddingLeft="@dimen/activity_horizontal_margin"
    android:paddingRight="@dimen/activity_horizontal_margin"
    android:paddingTop="@dimen/activity_vertical_margin"
    android:orientation="vertical"
    tools:context="com.example.xbg.multithreaddownload.MainActivity">
    <EditText
        android:layout_width="match_parent"
        android:layout_height="wrap_content"
        android:text="http://192.168.0.102/广播体操.mp4"
        android:id="@+id/FileUrl" />
    <ProgressBar
        style="?android:attr/progressBarStyleHorizontal"
        android:layout_width="match_parent"
        android:layout_height="wrap_content"
        android:id="@+id/progressBar" />
    <TextView
        android:text=""
        android:layout_width="match_parent"
        android:layout_height="wrap_content"
        android:gravity="right"
        android:id="@+id/tvProgress" />
    <Button
        android:text="下载"
        android:layout_width="match_parent"
        android:layout_height="wrap_content"
        android:id="@+id/btDown" />
</LinearLayout>
```

（4）本例需要访问网络，并将下载的文件保存到 SD 卡的公共目录 DownLoad 中，所以需要修改 AndroidManifest.xml，申请相应权限，代码如下。

```
<?xml version="1.0" encoding="utf-8"?>
<manifest xmlns:android="http://schemas.android.com/apk/res/android"
```

```
    package="com.example.xbg.multithreaddownload">
    <uses-permission android:name="android.permission.INTERNET"/>
    <uses-permission android:name="android.permission.WRITE_EXTERNAL_STORAGE"/>
    ...
</manifest>
```

（5）修改 MainActivity.java，代码如下。

```
package com.example.xbg.multithreaddownload;
import android.os.Environment;
...
public class MainActivity extends AppCompatActivity {
    private   int total;                    //记录当前下载量
    private int filelength;                 //记录下载文件的长度
    private boolean isDownloading;          //标识当前是否处于下载中，用于实现下载暂停功能

    //使用HttpURLConnection的getContentLength()方法获得的文件长度和实际的文件长度略有差异
    //所以在下载过程中，不能通过当前下载量和文件长度是否相等来判断下载是否完成
    //本例文件分3段在3个线程中下载，3个线程完成下载的时间不固定
    //所以用boolean数组来标记当前部分是否已经完成，只有3部分都完成后，整个文件才完成下载
    private boolean[] over;
    private   URL url;//记录下载文件的URL
    privateFilefile;//记录保存下载文件的File对象

    privateList<HashMap<String,Integer>>threadList;//保存每个线程所下载文件的开始、结束和完成量
    ProgressBarprogressBar=null;
    Button btDown=null;
    EditTextetFileUrl=null;
    TextViewtvProgress=null;
    @Override
    protected void onCreate(Bundle savedInstanceState) {
        super.onCreate(savedInstanceState);
        setContentView(R.layout.activity_main);
        initParameter();//初始化total、isDownloading等字段
        etFileUrl=(EditText)findViewById(R.id.FileUrl);
        progressBar=(ProgressBar)findViewById(R.id.progressBar);
        tvProgress=(TextView)findViewById(R.id.tvProgress);
        btDown=(Button)findViewById(R.id.btDown);
        btDown.setOnClickListener(new View.OnClickListener() {
            @Override
            public void onClick(View v) {
                //isDownloading为true，说明正在下载文件，此时单击按钮可暂停下载
                if(isDownloading){
                    isDownloading=false;//在子线程中会判断isDownloading，若为false则暂停下载
                    btDown.setText("下载");
                    return;
                }
```

```
                //isDownloading为false时，可能处于暂停或还没有开始下载的状态，此时单击按钮可继续下载
                //或者开始新的下载
                isDownloading=true;//设置标识，标识已经开始下载
                btDown.setText("暂停");
                new Thread(new Runnable() {//HTTP访问应该在子线程中执行，否则会抛出异常
                    @Override
                    public void run() {
                        doDownloading();//开始下载
                    }
                }).start();
            }
        });
        etFileUrl.addTextChangedListener(new TextWatcher() {
            //在EditText控件中修改下载文件的URL时重新初始化total、isDownloading等字段
            @Override
            public void afterTextChanged(Editable s) {
                initParameter();//在改变文件URL时执行初始化操作
            }
            @Override
            public void beforeTextChanged(CharSequence s, int start, int count, int after) {}
            @Override
            public void onTextChanged(CharSequence s, int start, int before, int count) {}
        });
    }
    private void doDownloading(){//将文件分段在子线程中下载
        if(threadList.size()==0) {
            //threadList为空时，说明还没有开始下载，所以从头开始新的下载
            try {
                String fileUrl=etFileUrl.getText().toString();//获得输入的下载文件URL地址
                url = new URL(fileUrl);
                HttpURLConnection conn = (HttpURLConnection) url.openConnection();
                conn.setRequestMethod("GET");
                conn.setConnectTimeout(5000);
                filelength = conn.getContentLength();//获得文件长度
                if (filelength< 0) {
                    Log.e("下载进度", "文件不存在！ ");
                    Toast.makeText(MainActivity.this, "文件不存在！ ",Toast.LENGTH_LONG).show();
                    return;
                }
                progressBar.setMax(filelength);//设置下载进度条最大值
                //使用进度条控件setProgress()方法设置当前进度时，可立即反映到界面的进度条控件中
                //虽然Android不允许在子线程中更新UI界面，但进度条控件setProgress()方法是个例外
                //所以不需要返回主线程中去执行setProgress()方法
                progressBar.setProgress(0);//设置进度条初始进度

                File path = Environment.getExternalStoragePublicDirectory(
```

```
                        Environment.DIRECTORY_DOWNLOADS);//获得SD卡的公共下载目录
            String fileName = fileUrl.substring(fileUrl.lastIndexOf("/")+1);      //获得文件名
            file = new File(path, fileName);                               //创建保存下载内容的SD卡文件
            RandomAccessFilerandomFile = new RandomAccessFile(file, "rw");
            int blockSize =filelength / 3;
            for (int i = 0; i < 3; i++) {//将文件分为3段下载
                int begin = i * blockSize;
                int end = (i + 1) * blockSize;
                if (i == 2) end = filelength;
                HashMap<String,Integer>map=new HashMap<String,Integer>();
                map.put("begin",begin);      //保存当前部分的开始下载位置
                map.put("end",end);          //保存当前部分的结束下载位置
                map.put("finished",0);       //保存当前部分的已下载量，初始为0
                threadList.add(map);
                //创建子线程下载文件，开始下载指定部分
                new Thread(new DownloadRunnable(i, begin, end, file, url)).start();
            }
        } catch (Exception e) {
            Toast.makeText(MainActivity.this, "出错了！",Toast.LENGTH_LONG).show();
            e.printStackTrace();
        }
    }else{
        //threadList不为空时，说明是从暂停状态继续下载的
        for(int i=0;i<threadList.size();i++){
            HashMap<String,Integer>map=threadList.get(i);
            int begin=map.get("begin");          //获得下载开始位置
            int end=map.get("end");              //获得下载结束位置
            int finished=map.get("finished");    //获得已下载量
            //创建子线程，继续暂停的下载，将i作为子线程ID
            new Thread(new DownloadRunnable(i, begin+finished, end, file, url)).start();
        }
    }
}
private void initParameter(){//初始化各个字段，在开始新的下载时调用该方法进行初始化
    isDownloading=false;
    over=new boolean[]{false,false,false};
    threadList=new ArrayList<>();
    btDown.setText("下载");
    tvProgress.setText("0%");
    total=0;
    filelength=0;
    progressBar.setProgress(0);
}
class DownloadRunnable implements Runnable{//下载子线程，实现文件分段下载
    private   int begin,end;         //记录文件的下载开始、结束位置
    privateFilefile;                 //保存下载内容的文件File对象
```

```
    private URL url;                    //保存下载文件的URL
    private int id;                     //保存子线程ID

    publicDownloadRunnable(int id,int begin, int end, Filefile, URL url) {
        this.begin = begin;
        this.end = end;
        this.file = file;
        this.url = url;
        this.id = id;
    }
    @Override
    public void run() {
        try {
            if(begin>end) return;//begin大于end，说明当前部分下载已经完成
            HttpURLConnection conn=(HttpURLConnection)url.openConnection();
            conn.setRequestMethod("GET");
            conn.setConnectTimeout(5000);
            conn.setRequestProperty("Range","bytes="+begin+"-"+end);//设置下载的文件范围
            InputStream is=conn.getInputStream();
            RandomAccessFile randomFile=new RandomAccessFile(file,"rw");
            randomFile.seek(begin);//设置保存下载内容的开始位置
            int len=0,finished;
            byte[] buf=new byte[1024*1024];
            HashMap<String,Integer>map=threadList.get(id);//根据线程ID来获得保存的下载信息
            while((len=is.read(buf))!=-1 && isDownloading){
                //当用户单击按钮暂停下载时，isDownloading变为false，当前下载停止
                randomFile.write(buf,0,len);
                finished=map.get("finished")+len;
                map.put("finished",map.get("finished")+len);//保存已下载量
                total+=len;
                progressBar.setProgress(total);//设置进度条下载进度
                runOnUiThread(new Runnable() {
                    @Override
                    public void run() {
                        //设置TextView中显示的百分比下载进度，因为需要更新UI，所以返回主线程执行
                        tvProgress.setText((int)(total*100.0/filelength)+"%");
                    }
                });
                if(end<=begin+finished) over[id]=true;//设置当前部分完成标识
            }
            is.close();
            randomFile.close();
            if(over[0] && over[1] && over[2]){//判断3个部分是否都已经完成
                Log.e("下载进度", "下载完成！");
                runOnUiThread(new Runnable() {
                    @Override
```

```
                            public void run() {//返回主线程执行，才能显示Toast和更新TextView文本
                                Toast.makeText(MainActivity.this,
                                                    "下载完成！",Toast.LENGTH_LONG).show();
                                tvProgress.setText("100%");
                            }
                        });
                    }
                } catch (Exception e) {
                    Toast.makeText(MainActivity.this,"出错了！", Toast.LENGTH_LONG).show();
                    e.printStackTrace();
                }
            }
        }
}
```

（6）运行项目，测试运行效果。

在使用真机进行调试时，将应用安装到设备后，需要在系统的设置管理中为应用授予存储访问权限，否则会无法将下载的文件保存到 SD 卡中。

8.4 小结

通常，一个应用程序至少有一个进程，一个进程至少有一个线程。进程拥有独立的运行内存，而多个线程则可共享内存，从而极大地提高程序的运行效率。从逻辑上看，多个线程可以同时运行。线程不能够独立执行，必须依存应用程序进程，由应用程序提供多个线程执行控制。在实现比较耗时的任务时，使用多个线程可以提高任务的完成效率。

服务通常在后台运行，完成后台任务。在服务中也可使用线程，两者并不冲突。

8.5 习题

1. 请说明如何创建线程。
2. 请问能够在子线程中改变界面中 TextView 控件的文本吗？
3. 请问 AsyncTask 的哪些方法在主线程中执行？哪些方法在子线程中执行？
4. 请问如何定义一个服务类？